Wales

The reconstructed bridge with the new road above the railway.

CIVIL ENGINEERING HERITAGE

Wales

Keith Thomas

PHILLIMORE

2010
Phillimore & Co. Ltd
Andover, Hampshire, England

ISBN 978-1-86077-638-0

ALSO IN THE CIVIL ENGINEERING HERITAGE SERIES
(General Editor: Peter Cross-Rudkin)

East Anglia by Peter Cross-Rudkin

West Midlands by Roger Cragg

CONTENTS

PREFACE

Britain has a heritage of civil engineering structures and works unrivalled anywhere else. The skills of past engineers are in evidence throughout the land in a fascinating variety of bridges, structures, utilities and lines of communication: the infrastructure of society.

The series *Civil Engineering Heritage* makes available information about these works to a wide public in order to broaden the interest and understanding of the reader in the immense range of expertise that features in our history. In recent years people have become more aware of the value of our heritage and our responsibility to keep the best examples for future generations. It is therefore hoped that these books will assist those seeking to advance the cause of conservation.

Much of the information in these books, particularly in the gazetteer sections, has been extracted from the records of civil engineering works prepared by the Institution of Civil Engineers Panel for Historical Engineering Works (PHEW). The works covered by these records have been selected for their technical interest, innovation, association with eminent engineers or contractors, rarity or visual attraction. The Institution's Archives have become the principal national repository for records of civil engineering and are now regarded as the leading authority on historical engineering works. They are widely consulted by heritage organisations, the business community and those involved in private research. The records may be consulted by the public for research purposes on application to the Institution's Archivist. The relevant Historical Engineering Work number is given, where one exists, for each item in the gazetteer to assist researchers in following up items of interest. Additionally, PHEW has a number of subpanels dealing with specific types of civil engineering works in greater detail, who are able to assist with information about works within their remit.

ACKNOWLEDGEMENTS

I am grateful for the assistance, help and advice I have received in preparing this volume. I am particularly grateful to my support team in Wales, Bob Daimond, Stephen K. Jones and Tony White, for revisiting and updating the information on the HEW records. My thanks also to Roger Cragg, the editor of the second edition of the earlier heritage book *Wales and West Central England*, for the re-use of much of his material for the gazetteer; to past members of the Panel in Wales for their considerable efforts in assembling many of the current records; to the late Paul Dunkerley, Northwest Association Panel Member, who did much of the work in North Wales prior to it being incorporated in an all-Wales Association.

I acknowledge the additional help and advice that I have received from other civil engineers in Wales who have supplemented my knowledge of some specialist areas, notably David Evans and Philip Marsden. Help and advice have also been received from Barry Wilkinson, Chairman of the IGEM History Panel and editor of *Historic Gas Times*; Brian Malaws of the Welsh Mills Society; Trinity House, Swansea; Barry Barton; Professor Ben Barr; and Professor Tim Hughes. Thanks also to Peter Cross-Rudkin for advice and suggestions; and not forgetting my wife, Vivienne, for her help with the typing, advice and accompanying me on my travels to check details of the text and to photograph sites. All errors and omissions, however, remain mine.

Photographs have been obtained from various sources and I am grateful to Roger Cragg, Stephen K. Jones, Tony White, Bob Daimond, Nick Holyfield of Canolfan Thomas Telford Centre, Menai Bridge, Peter Cross-Rudkin, William Day and Philip Banks for the use of their photos; to Carol Morgan and the staff of the Institution Library for finding items from the Institution's collections for me; to the Society for the Protection of Ancient Buildings for the use of photographs from Jervoise's 1936 book on the ancient bridges of Wales; to the Director of Transport at the National Assembly for Wales; to Stewart Williams for the use of photos from his series of books *Cardiff Yesterday*; and to Bob Slorach of Associated British Ports. Additional material has been supplied by Dr Noel Meeke of the Waterworks Museum, Hereford; Milford Haven Port Authority; Bryn Williams of First Hydro; the National Slate Museum, Llanberis; and Nigel Young of Newport Past. Copyright of these items remains with those named. Some photographs are from the collection of the late Owen Gibbs, my predecessor as Panel Member, who passed to me his extensive collection of photos and slides. All except my own have been attributed to the provider.

I am grateful to the Royal Commission for Ancient and Historic Monuments in Wales (RCAHMW) and the National Library of Wales (NLW) for the use of items from their collections. Aerial views of Barry and Holyhead are Crown Copyright RCAHMW – Aerofilm Collection; other items are Crown Copyright RCAHMW, except Fig. 59, which is from the collections of the National Monuments Record of Wales. Items from NLW are by permission of NLW. My thanks also to Heather Robbins of Phillimore for her work in finalising the layout of my text and photographs.

Civil engineers at the heart of society, delivering sustainable development through knowledge, skills and professional expertise.

Peirianwyr sifil yng nghanol y gymdeithas, yn cyflawni datblygiadau cynaliadwy drwy wybodaeth, sgiliau ac arbenigedd proffesiynol.

INTRODUCTION

Civil engineering has for centuries made a major contribution to civilisation in providing the infrastructure that enables our daily lives. Civil engineers, even before they were known as such, have played a dominant role in the provision of transport infrastructure, roads, canals, railways, ports, harbours and airports; in public health in the provision of water supply, drainage and sewerage; coastal protection, land reclamation and drainage; in power supply from water, wind and fossil fuels; and construction of industrial facilities, offices and housing.

Britain has been in the forefront of engineering and innovation for many years. The historical engineering works listed in this book are examples from Wales of the skills of past and the present civil engineers. It is hoped that both those with an engineering background and the more general reader will gain a better understanding of the diversity of expertise and experience that has contributed to our civil engineering heritage and be encouraged to see for themselves these examples of the work of the profession around us.

The profession of civil engineer only became a recognised description in the mid-18th century and a formal designation with the formation of the Institution of Civil Engineers in 1818. The Society of Civil Engineers was formed in 1771 by John Smeaton and a group of other leading engineers of the time. He was the first self-proclaimed civil engineer, coining the term to differentiate his work from that of the military engineer. The society, the first engineering society in the world, became a meeting place for the leading civil engineers of the time but did not satisfy the needs of younger engineers in developing their knowledge and training. Hence a small group founded the Institution of Civil Engineers, which received its Royal Charter in 1828 as the world's first professional engineering body.

Early civil engineers often received their training as millwrights, carpenters and masons. Much of their design work was empirical and based on experience rather than mathematical calculation. As structures became more complex the need for a greater understanding of the design principles became necessary.

Wales is said to have been one of the first industrialised nations in the world, when the number of people employed in industry exceeded those on the land. In the latter half of the 18th and the early 19th centuries industrialisation spread rapidly, particularly in the south and the north east. The rivers and sea had been the main transport routes for centuries, land communications being poor and unreliable. The new industries required better transport links for the movement of raw materials and finished goods. The need for efficient low-cost land transport systems began with the development of the canal network and horse-drawn tramways, followed by the intense development of ports and railways. As towns grew the need for building materials, water supply and drainage also grew rapidly. All this employed the ingenuity of many of Britain's finest engineers and the names of many will be familiar, including James Brindley, William Jessop, the Dadfords, John Rennie, Thomas Telford, Robert Stephenson, Isambard K. Brunel and the Welsh-based William Edwards, Watkin George, George Overton and L.G. Mouchel. There are also many lesser-known engineers, all of whom have made their contribution to the development of civil engineering, not forgetting the builders and contractors who undertook these works.

The achievements of the 18th- and early 19th-century engineers are even more remarkable when it is remembered the equipment and communications available to them were considerably different to today. Journeys we now measure in hours would take days, and telephone and internet connections were of course non-existent. While documents and to a certain extent drawings could be reproduced by printing, written communication was still by hand, a copy being kept in letter books. Replies could take days to be received. Very little construction machinery was available, mainly medieval hand-operated methods of cranes, derricks, winches and hydraulic jacks. Transport was horse-powered: one horse on a canal could haul about 20 tons. On a tramway one horse on level ground could haul about eight tons, four horse-power being about the practical maximum. Excavation equipment was similarly basic. The canal navvies dug by hand and moved materials by barrow and horse and cart. The development of powered construction equipment from the early 19th century is a hidden achievement of the engineering construction industry behind many of the projects described in this book. With all this in mind the achievements of the designers, consultants and contractors of the time and the workmen themselves, of whom we know very little, are even more worthy of recognition

This book covers the whole of the Principality with its diverse history and topography and identifies sites of historical civil engineering interest. The area is dominated by its major rivers which have been both a means of communication and a barrier to it. Many of the works described in this book have arisen from these two aspects.

The first part has been divided into chapters by specialist subject but there is of course considerable overlap and linkage between them. It explains briefly the historical development of different facets of civil engineering with references to example sites to illustrate the subject. This is followed by a gazetteer which gives brief descriptions of typical historical engineering works. For sites which are in the gazetteer the gazetteer reference number is given. Sites which are not in the gazetteer are identified by their national grid reference. To assist anyone wishing to visit sites in the gazetteer the location of each is marked on maps at the head of each section.

The spelling of some Welsh place-names has changed over the years. Where the old spelling was used on documents or details at the time this has usually been retained. Elsewhere the current spelling has been used (for example Llanelly is now Llanelli but the original spelling has been retained for the Llanelly Railway).

Most of the material in the gazetteer has been abstracted from the Historical Engineering Works (HEW) Records, which have been prepared over the past few decades by Panel Members and their helpers. Each work has an HEW number and in the gazetteer this together with a grid reference is given for each site. Not all HEWs in Wales are described for reasons of space. An index of engineers and contractors is also included.

This book does not set out to be an authorative history of the engineering infrastructure of the Principality; there are many specialist books which comprehensively cover each aspect of industrial and transport infrastructure. What it aims to do is identify extant features to illustrate that history and show how they link into the development of civil engineering design and construction over many years.

Keith Thomas, CBE, BSc, C. Eng, FICE, FCIHT

METRIC EQUIVALENTS

Imperial measurements have been generally adopted for the dimensions of the works described in this book as this would have been the system used at the time of the design and construction of most of them. Where modern structures have been included which were designed and built using the metric system, metric units have been used in the text.

The following are the metric equivalents of the imperial units used:

Length:	1 inch = 25.4 millimetres
	1 foot = 0.3048 metres
	1 yard = 0.9144 metres
	1 mile = 1.609 kilometres
Area:	1 square inch = 45.2 square millimetres
	1 square foot = .0929 square metres
	1 acre = 0.4047 hectares
	1 square mile = 259 hectares
Volume:	1 gallon = 4.546 litres
	1 million gallons = 4546 cubic metres
	1 cubic yard = 0.7646 cubic metres
Mass:	1 pound = 0.4536 kilograms
	1 ton = 1.016 tonnes
Power:	1 horsepower (h.p.) = 0.7457 kilowatts
Pressure:	1 pound force per square inch = 0.06895 bar

INLAND WATERWAYS AND EARLY TRAMWAYS

The use of rivers and coastal waters for navigation, communication and trade in Britain dates from the earliest times, certainly the pre-Roman era. After the end of the Roman occupation those roads that existed generally fell into a state of disrepair and rivers and the sea continued to provide an alternative means of transport for freight and passengers. The major rivers such as the Severn were important arteries of trade and it is no surprise to find that early industrial developments often took place along their banks. The development of the ironfounding industry at Coalbrookdale on the River Severn is a typical example since it had both a convenient source of raw materials (coal and ironstone) and the river as a convenient transport system. The Severn Estuary provided a major trading and transport route between Bristol, south-west England and South Wales as late as the 19th century with regular packet sailings to various ports and harbours along the coast. Cross-Severn steam packets flourished and were used regularly by the travelling public, including Brunel, who on occasion used the Bristol to Cardiff packet to visit his project on the Taff Vale Railway and potential suppliers in South Wales for his other engineering projects.

The golden age of this network was in the period before the canals and railways, when all possible use was made of coastal transport and manufacturers went to great lengths to transport their goods. Jack Simmons (1962) writes in his book *Transport: A Visual History of Modern Britain (p.32)* that for example in 1775 the Horsehay Company of Wellington near Shropshire was sending pig-iron to Chester by river and sea, a journey that began with transport by cart to the Severn. It would be loaded on to Bristol-bound riverboats and there trans-shipped to vessels sailing around the coast of Wales and up the Dee to arrive at Chester. Transport by this route involved a journey of over 400 miles by sea and two trans-shipments, compared with 60 miles by land, and well illustrates the state of the roads and the reluctance of carriers to use them. The River Severn was navigable, when river levels permitted, as far as Pool Quay above Welshpool, over 100 miles from its estuary. Here oak from the forests of Montgomeryshire was carted to a wharf on the river, where it was loaded onto ships or formed into rafts and floated down the river to be loaded onto larger vessels. General cargoes were brought in and flannel, non-metallic ores and limestone were loaded as return cargoes. The vessels would be hauled by manpower along the river bank when winds were unfavourable to sail. The main Welsh river used for transport was the Wye on the English border from Chepstow to Monmouth (and to Hereford).

Rivers were also a convenient source of power for industry and early attempts to improve the navigation of rivers often gave rise to conflicts between the constructors and the mill owners, who regarded navigation as a threat to their water supplies. However, rivers were successfully improved by the provision of weirs to increase river levels, with flash or pound locks to allow boats to pass the weirs, and in some cases lengths of artificial waterway were constructed to bypass difficult lengths of river. Flash locks were simple single gates set in the weir which when opened allowed a boat to pass downstream on the 'flash' of water. To go upstream the boat would be hauled against the water flow. Pound locks were the more familiar arrangement of an impounded lock between two sets of gates, which was less wasteful of water but more difficult to construct, requiring a sluice arrangement for filling or emptying each time a boat passed through.

Purely artificial waterways (canals) date from Roman times with the building of the Fossdyke from Torksey on the River Trent to Lincoln, a distance of 11 miles. However, the large-scale building of canals in the British Isles dates from the mid-18th century, starting with the cutting of the 18½-mile Newry Canal in Northern Ireland in 1742, also the first summit-level canal (a summit canal being one that falls both ways from a high point rather than one that starts at a high level and descends only). This was soon followed by the Sankey Brook navigation in 1757 and the Duke of Bridgewater's canal from his coal mine in Worsley to Manchester in 1761. Following the success of these early canals other waterways were rapidly promoted, including the Trent & Mersey, the Staffordshire & Worcestershire, the Coventry and the Oxford Canals. These canals linked the four major rivers of England, the Trent, Mersey, Severn and Thames, and thus provided a more reliable freight transport system for the nation. Many of these early canals were built under the supervision of the pioneering canal engineer James Brindley.

The commercial success of these early canal companies sparked a massive increase in canal promotion, the period from about 1790 to 1797 seeing the authorisation of 53 new canals and many more schemes which never reached the stage of parliamentary sanction. This period is often referred to as the 'Canal Mania era'. Inevitably, many of the canals built during this period were not commercially successful. Many miles of canal were built to satisfy the increasing demand for freight transport, a consequence of industrialisation in the latter half of the 18th century. Canal revenue was from tolls charged for the use of the waterway by the boat owners. There was one canal in Wales, the Tennant, which did offer a horse-towing service, but this was very much the exception.

From the 1830s the widespread development of railways inevitably meant that the inland waterway network suffered from the competition. Many canal and river navigation companies passed into the hands of railway companies. A slow decline of traffic and revenue took place with consequent lack of maintenance leading to further loss of traffic. However, much of the national network of inland waterways remained intact, one notable example being the intensive network of canals in the Black Country around Birmingham where the canals were used as a local distribution system for freight carried into and out of the area by the railways.

Following the Second World War, freight-carrying on waterways declined rapidly but the system has undergone a remarkable transformation in recent years, becoming a major leisure facility with increasing numbers of privately owned and hire craft using it each year and in some places a resurgence of commercial use.

THE CANAL SYSTEMS IN WALES

In Wales canal development took different forms in the North and South. In North and Mid Wales canals were extensions of English systems to transport materials to markets and businesses in England. In particular the transport of lime for the improvement of farmland was important. Canals were also used as a source of water supply to industries further afield. This aspect has helped with the retention of some parts of the system even after closure to commercial traffic. For example the Llangollen Branch of the Ellesmere Canal provides a supply of up to six million gallons a day (mgd) to a reservoir at Hurleston in Cheshire. This was originally built in the 1830s and enlarged in the 1950s for that purpose.

As the use of coal developed the transport of this and other materials developed as well. Industries grew up alongside the canals because of the convenience of transport and availability of materials as well as a supply of water.

North and Mid Wales

Both the **Llangollen Branch (N37)** of the Ellesmere Canal and the **Montgomeryshire Canal (M6)** became branches of the Shropshire Union canal network. The Llangollen Branch, which was built between 1795 and 1805 by William Jessop as consulting engineer and Thomas Telford as general agent and Engineer, includes Telford's world-famous **Pontcysyllte Aqueduct (N39)**; one of the largest earthworks of the time in the southern approach to the aqueduct; **Chirk Tunnel (N41)** (one of three tunnels on this branch all distinguished by including the towpath alongside the canal); **Chirk Aqueduct (N40)**; and, west of Llangollen, **Horseshoe Falls Weir (N38)** on the River Dee, which provides the water for the canal. This branch of the canal system supplied coal, clay, limestone and ironstone to industries alongside it in Wales and in England. It has remained in use when others have declined and has now developed into a major leisure facility and tourist attraction. The original plan was for the canal to extend to Chester to serve the growing industries around Ruabon and Wrexham and to link with the existing canal system and the River Mersey, with the channel to Llangollen only constructed to supply water for the canal, but in the event this extension was never built and water supply became the dominant function. In June 2009, 11 miles of the Llangollen Canal, including Pontcysyllte Aqueduct, were granted World Heritage Status in recognition of the civil engineering achievements and innovation of Telford and Jessop.

In North Wales ironworks were developing along the canal. Iron ore was available in the Wrexham area as well as coal for coke, limestone and water power. In 1717 Charles Lloyd built a furnace at Chirk, and in the 1730s ironmasters in Bersham near Wrexham were making small cast-iron goods. In 1753 Isaac Wilkinson, a north-country ironmaster, bought the works and expanded it to make larger items including armaments. His son, John 'Iron Mad Jack' Wilkinson, took over and became a leading figure of the time. He developed a boring machine that made accurate cylinders which were in great demand for the new steam engines and his cannon-boring machine supplied the Royal Navy with the most accurate cannon in the world. He patented a steam rolling mill and a cupola furnace for high-quality castings. He built a new ironworks at Brymbo near Wrexham in 1793 and added additional furnaces in 1805. The Bersham works closed in 1812. Brymbo was one of the first companies to exploit the new Bessemer process to make steel in 1856. Other ironworks developed in the area, notably William Hazledine's works at Plas Kynaston near Ruabon, built to supply the ironwork for Pontcysyllte Aqueduct. It subsequently supplied ironwork for several of Telford's later bridges, including one as far away as Craigellachie near Inverness.

The Eastern Branch of the **Montgomeryshire Canal (M6)**, just over 16 miles long and designed by John Dadford, was built between 1794 and 1797 from a 10-mile branch of the Ellesmere Canal from Frankton to Llanymynech (Carreghofa) on the Welsh border. This extended the canal to Garthmyl to bring lime to England from Llanymynech and Porthywaen quarries. It also brought coal from the coalfields at Chirk and Ruabon to the many limekilns established along its route and transported high-quality oak and elm from the county, timber that was especially in demand by the Royal Navy for shipbuilding. There were over 90 lime kilns along the route and 13 specialist timber wharves. Other trade developed, including flannel and woollen production, ironworks and granaries. The canal remained predominately agriculturally based with many mills and timber yards using its available water supply to drive their machinery.

John Dadford left in 1796 and the work was taken over by his father, Thomas Dadford senior. Between 1815 and 1819 it was extended a further seven miles by the Western Branch from Garthmyl to Newtown, with Josias Jessop as consulting engineer. His father, William, had been consulting engineer for the Ellesmere Canal. In 1847 the Eastern Branch was taken over by the Shropshire Union Canal Company and in 1850 the Western Branch was also transferred. The canal was disused by 1936 and abandoned in 1944 but restoration work continues by a very active preservation society. Over 17 miles of the 35-mile length from Frankton have been reopened for leisure use in two sections. One is linked to the main line at Frankton and the other is at Welshpool. Most of the rest is in water with only one dry section in England where work continues on restoration. However, in several locations outside these navigable sections the canal has been bridged by road improvements at canal level, preventing through-navigation.

1 *Milepost at Vyrnwy Aqueduct.*

Significant features on the Eastern Section include Dadford's four-span masonry **Vyrnwy Aqueduct (M6.1)** over the river at Llandysilio, which is awaiting restoration. **Brithdir Aqueduct (M6.3)** over the Lower Luggy at Berriew is in use. **Berriew Aqueduct (M6.4)**, another Dadford four-span masonry construction over the River Rhiw, had major rebuilding in 1889. It has been restored and is open to navigation. **Penarth Weir (M6.5)** at the head of the canal at Newtown provides the water supply for the Western section. There are many opportunities for walkers along the towpaths.

West and South Wales

There was little inland navigation in South and West Wales. Many rivers had tidal wharfs but these were close to the river mouth; access further upstream was dependent on tide and very variable. The earliest Welsh canals were constructed in West Wales but were merely short lengths linking collieries to shipping quays on the rivers. None were more than a few miles long. Probably the first was a short canal on the River Neath with the first sea lock in Wales, built around 1700 by Sir Humphrey Mackworth, but it was little more than a basin with tidal gates. The River Neath had been navigable as far as Aberdulais and several such small improvements had been made. The river trade was important for the movement of raw materials and finished products. At Aberdulais copper smelting is known as early as 1584 and iron working in 1667.

The first authentic canal was Thomas Kymer's Canal, *c.*1769, about three miles long, bringing coal and lime from his colliery and lime works at Pwll Llygod to the coast at Kidwelly. Over the next 30 years several more short canals were built for improved access to river wharves from Neath westwards as far as Kidwelly and Burry Port.

In South Wales the development of ironworks at the heads of the valleys where ironstone, timber (for charcoal) and limestone were readily available meant that a transport system was needed to move the finished products south to the markets. The iron was transported by packhorse, often on new turnpikes constructed by the ironmasters. The Merthyr Tydfil to Cardiff turnpike opened in 1771. The industry grew from the 1760s, utilising coal available close to the ironstone and limestone at the heads of the valleys as supplies of timber dwindled, using methods developed by Abraham Darby at Coalbrookdale. This method of transportation was completely inadequate. Something new was needed. Canals were being constructed in England and the idea was readily accepted by the ironmasters.

Canal development took a different form here as a consequence of the north-south alignment of the valleys. Each valley had its own individual canal and only in two instances was a canal link between systems practical. The Brecknock & Abergavenny Canal in the Usk Valley linked with the Monmouthshire Canal (**S11**) in the Usk and Ebbw Valleys. The

Tennant Canal (S34.2) linked to the **Neath Canal (S34.1)** and provided a route between the Neath and Tawe Rivers.

Between 1794 and 1799 four major canals were built. The engineers were Thomas Dadford, senior and junior, and Thomas Sheasby, also senior and junior, who had developed their expertise building canals in the Midlands. The first to be completed was the Glamorganshire Canal from Merthyr Tydfil to Cardiff, authorised in 1790 and opened in 1794; followed by the Neath Canal, 1795; the Monmouthshire Canal with two lines from Newport to Crumlin and to Pontypool, 1796; and the Swansea Canal in the Tawe Valley in 1798. The Glamorganshire Canal added a sea lock pound in 1798; the others used river wharfs. The Neath Canal was extended southwards to Giants Grave near Briton Ferry in 1799 to an improved river wharf. (An impounded dock designed by Brunel was built further downriver between 1854 and 1861.) In all, 77 miles of canal and 180 locks were built in this period.

The number of major ironworks at the heads of the valleys between Hirwaun at the western end and Blaenavon at the east grew from three in 1780 to eight by 1790, and more followed. By 1815 there were over 100 furnaces in operation. As an illustration of the rate of growth the four ironworks in Merthyr – Cyfarthfa, Plymouth, Penydarren and Dowlais – produced 16,000 tons of iron in 1792; this rose to 43,000 tons by 1819. Blaenavon produced 1,000 tons in 1801 and 38,000 tons in 1819. In 1845 the Dowlais Ironworks at Merthyr Tydfil was said to be the biggest in the world with 18 furnaces, seven more than Cyfarthfa and 12 more than Penydarren. It was producing 88,000 tons, Cyfarthfa about 46,000 tons and Penydarren 16,000 tons.

The canals formed a backbone to an associated network of tramways bringing materials to them and to the ironworks. The topography of the South Wales Valleys precluded branch canals except for a few short sections on the coastal plain and tramways provided the access to and from mines, quarries and foundries. Tramways could be constructed up to eight miles from the canal without further parliamentary approval (the maximum distance empowered by the Canal Acts for supplementary tramroads). One proposed branch canal along the foreshore at Swansea to Oystermouth was eventually built as a tramway after local objections. Tramways were horse-drawn as were the canals until the development of steam locomotives, and some remained horse-drawn into the mid-1800s.

2 *Commemorative plinth at the start of the Merthyr (Penydarren) Tramroad in Merthyr Tydfil.*

The **Merthyr Tramroad (S25)** was built between 1799 and 1802 from Penydarren in Merthyr Tydfil to the Glamorganshire Canal at Abercynon by a group of ironmasters on the east side of the River Taff in Merthyr Tydfil. The canal above Abercynon had a water supply problem and priority of use was given to another Merthyr iron master for his ironworks at Cyfarthfa on the west side of the Taff. It was on this tramroad that the world's first steam locomotive ran when in 1804 Richard Trevithick successfully demonstrated his new railway engine. It was not, however, until the construction of the Taff Vale Railway, opened between Cardiff and Abercynon in 1840 and completed to Merthyr Tydfil in 1841, that steam locomotion began to replace the horse in South Wales.

The growth in the use of coal as a fuel also led to the development of more collieries and tramroads and later new railways as the demand grew and production increased. By 1830 there were 350 miles of tramways in South and West Wales.

3 *Commemorative plaque at the end of the Merthyr Tramroad at Abercynon.*

Later canals include the Brecknock & Abergavenny Canal, which although begun in 1793 was not completed until 1812, when it linked southward from Llanfoist to the Monmouthshire Canal at Pontymoile north of Pontypool. It was built to bring coal from the Clydach Valley to wharves at Gilwern and Llanfoist for the Usk valley and to serve ironworks at Clydach and Beaufort. The Aberdare Canal was built from Aberdare to connect with the Glamorganshire Canal at Abercynon in 1812 to serve ironworks in Aberdare and Hirwaun. The Kidwelly & Llanelly Canal was commenced in 1812 inland to Pontyberem and to connect with Kymer's Canal. The Tennant Canal was built from 1817, incorporating several small earlier canals to access the River Neath at Red Jacket Pill and connecting to a new harbour, Port Tennant Tidal Harbour, on the Tawe at Swansea (later in 1881 becoming part of the Prince of Wales Dock, Swansea, as an impounded dock). It was extended in 1824 to link with the Neath Canal at Aberdulais.

The growth in traffic was substantial. In 1819 South Wales canals carried about 80,000 tons; by 1829 this had risen to 192,000 tons and by 1839 to 372,000 tons. By the 1830s there was an inward trade in iron ore as local ores were used up and in timber for pit props, as well as domestic goods and foodstuffs.

Apart from the Llangollen Canal and the Monmouthshire & Brecon Canal there are no substantial lengths of canal still in use in Wales. Short sections of the Montgomeryshire Canal, the Monmouthshire Canal branch to Crumlin and the Neath and Tennant Canal have been restored by canal restoration societies and work continues on re-establishing further lengths of these canals.

For the Glamorganshire Canal there is little that remains. Here and there individual features may be found but much has been buried under new roads in the Taff Valley. At Merthyr Tydfil in front of Chapel Row and Joseph Parry's cottage a reconstructed canal section complete with a typical cast-iron canal bridge relocated from Rhydycar can be seen (SO 041071).

4 *The relocated Rhydycar Bridge at Merthyr Tydfil.*

A short derelict section also still exists to the east of the now demolished Brown Lenox Chain Works at Pontypridd (ST 080900). Chains were loaded at a wharf here for shipment via Cardiff. At Melingriffith in North Cardiff a short section now forms part of a nature reserve and here is the reconstructed **Melingriffith Water Pump (S17)** (ST 143800), which raised water from the Taff via a feeder to the canal until the canal closed in 1942. It was restored in 1974.

At the south end of this section at Melingriffith is an interesting towpath bridge, which crosses an overflow from the canal back to the River Taff (ST 145805). The bridge has a curved parapet on the canal side of the towpath to avoid snagging the towing rope as boats were pulled past.

5 *Bridge at Melingriffith, Whitchurch.*

The Aberdare Canal was built to the Glamorganshire Canal at Abercynon. A section at the top end may still be seen at Canal Head, Aberdare (SO 018024).

The Monmouthshire Canal has a considerable navigable length north of Newport to Pontymoile and on to the Brecknock & Abergavenny Canal. There are many original features over this whole stretch, canal bridges and aqueducts. **Brynich Aqueduct (S12)** just south of Brecon carries the canal at high level over the River Usk. The Crumlin Branch is partially restored but the long staircase of **Fourteen Locks (S13)** at Risca still has to be reinstated. The lock chambers have been cleared of debris and the uppermost lock rebuilt. The Fourteen Locks Visitor Centre is located alongside the uppermost lock.

The **Neath Canal (S34.1)** has also been partially restored and several miles are in use above Resolven. A section from the town of Neath south to Giant's Grave is also being restored by an active canal preservation group.

At Ynysbwllog (SS 803011) on the Neath Canal a three-arch masonry aqueduct was partially washed away in a flood in 1980. As the canal was in use as a water supply, temporary pipes were laid over the breach. The aqueduct has now been replaced with a new one spanning the river. This is said to be now the longest single-span aqueduct in Britain. The photos show the aqueduct in 1977 before the flood damage, the aqueduct after the water supply had been restored by pipelines and the new aqueduct built in 2008.

6 *Fourteen Locks – upper lock and basin. The Visitor Centre is on the left of the picture.*

Anti-clockwise from top left:

7 *Ynysbwllog Aqueduct, 1978. (© PC-R)*

8 *Ynysbwllog Aqueduct in 1980 after flood damage. (© PC-R)*

9 *New Ynysbwllog Aqueduct, 2009.*

10 *The six-arch masonry aqueduct at Glastony is now a scheduled ancient monument.*

11 *Swansea Canal locks at Trebanos.*

A substantial length of the **Tennant Canal (S34.2)** still exists. The main features of this canal are at its junction with the Neath Canal at Aberdulais. Here is its only lock, complete with lock-keeper's cottage and tollhouse and a substantial 10-arch aqueduct over the river to the connection with the **Neath Canal (S34.1)** (SS 774994).

There is little of the Swansea Canal remaining, but the length from Clydach just north of Swansea to Pontardawe and a section north of Pontardawe still exist. On these sections remains of the locks are visible, converted into weirs to maintain the water flow. **Twrch Aqueduct (S41)** on a derelict section at Ystalyfera has been restored. The canal has virtually disappeared in this area but the aqueduct has been retained (although without water) as a footpath and cycleway. This aqueduct was lined with hydraulic lime concrete, the first recorded use of this material for this purpose.

The Kidwelly & Llanelly Canal was opened in 1815 and extended in 1839. The **Burry Port & Gwendraeth Valley Railway (S38.1)** was built over and alongside it in 1869. The railway re-used canal structures and the notable one is the converted aqueduct at Glanstony, Trimsaran (**S39**) built in 1815.

TRAMROADS AND EARLY RAILWAYS

Tramroads

The first coal mining and metal smelting in Wales occurred where the coalfield was near the sea at Neath and Swansea. The earliest tramroads recorded in Wales are Mackworth's late in the 17th century at Neath and Morris's in the mid- to late 18th century north-west of Swansea. These were of narrow gauge as the waggons went underground into the mine workings with limited clearance. In 1756 a Newcastle engineer, George Kirkhouse, replaced by Cumbrian engineer Edward Martin in 1783, was employed at the Llansamlet Coalfield to construct over 12 miles of wooden waggonway both above and underground. This line from a coal pit at Landore also had one of the earliest surface railway tunnels to a riverside quay at Swansea.

As the iron and coal industries expanded in South Wales, traffic on the canals grew and congestion followed. Water supply problems servicing the many locks, icing and frost in winter, combined with the growing volume of traffic meant alternative solutions were needed. The **Merthyr Tramroad (S25)** described earlier is an example of a partial solution bypassing a congested section of the Glamorganshire Canal above Abercynon. This is one of the few tramroads that were not later developed for rail or road use. Most of its length can still be travelled on foot or cycle. Part is a section of the Taff Trail and the remainder south of the Trevithick Tunnel in Merthyr Tydfil has been restored by the Local Authority as part of the Trevithick 200th anniversary in 2004. In many places the original stone blocks that supported the rails can still be seen. By 1840 the tramroad was carrying over 75,000 tons of traffic.

Elsewhere in South Wales existing canal tramroads were extended southwards and in some cases paralleled a whole canal or used the original canal structures as

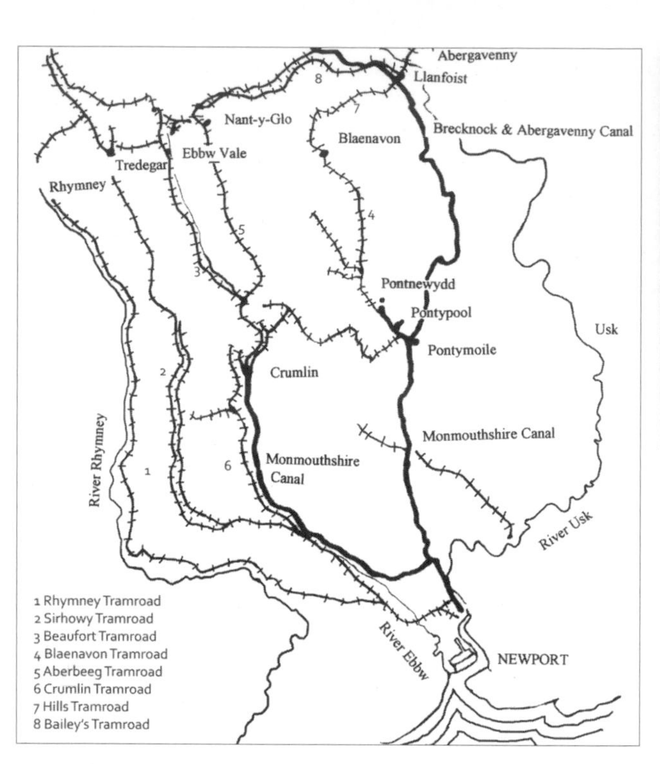

13 *Tramroads of Monmouthshire.*

12 *Stone sleeper blocks near Quakers Yard.*

part of the route. The Eastern and Western Valleys above Newport in Monmouthshire saw a proliferation of tramroads to service the canals, initially to access the canals since the topography militated against canal building, but later they developed into a railway network.

One tramroad developed in this period was not, however, to relieve canal congestion. The Brecon Forest Tramroad (SN 807122 to SN 917283), almost 26 miles long, was built between 1822 and 1825 by the London-based Scottish merchant John Christie to transport lime from his quarries and kilns at Penwyllt at the head of the Swansea Valley north to his newly acquired lands on the uplands of Breconshire and to those of his neighbours. The government had sold upland areas in England and Wales to help fund the Napoleonic wars. The land was very acidic and needed large quantities of lime to improve its fertility. The line also brought coal from his mines in the upper Swansea valley to his kilns to produce the lime. Christie's agent and surveyor was David Jeffries of Ysclydach, his engineer Joseph Jones of Ystradgynlais. To improve the profitability of the enterprise he also incorporated an existing short link to the Swansea Canal to export coal; at one time Christie owned three ships operating out of Swansea and 24 canal boats, representing a quarter of the canal's trade. New farms were built and connected to the northern end of the tramroad at Sennybridge as well as seven miles of new road and 44 miles of walls to surround his land, but in 1827 the overall enterprise of the tramroad and the development of the upland areas bankrupted him.

Ownership of the tramroad passed to the Claypon brothers, London bankers and creditors of Christie. Joseph Claypon built another link to the Swansea Canal at Gurnos for the new ironworks at Ynysgedwyn, Ystradgynlais and, later, Ystalyfera. Iron production expanded rapidly from three furnaces in 1837 to 36 by 1848, 22 adjacent to the tramroad. Ystalyfera, established in 1838, had the second largest furnace bank in South Wales and at one time was the largest tinplate works in the world. Total production peaked in 1857 at just over 64,000 tons. The engineer for the tramroad extension and further branches to the original line of about 12 miles was David Thomas, until his departure to the United States in 1838 leaving the main role to the well-known engineer of the period William Brunton. Brunton was involved with work for the Ynysgedwyn Ironworks and links to the main tramroad. The tramroad eventually closed in 1863. The terminus buildings of the tramroad at Castell-du, Sennybridge, still exist as a farm complex but much of the route has been taken by the later railway and road. Parts of the route can be traced and some of the structures are still visible.

Some early bridges on other tramroads can still be seen and those of particular interest are described in the bridges section later.

Early Railways

The change from tramroads to railways has no firm transition date, as the process was gradual. The world's first railway locomotive, mentioned above, was designed and built by Richard Trevithick for the Merthyr Tramroad in 1804. He developed this from his high-pressure static steam engines used in the Cornish tin mines, themselves an improvement on James Watt's development of the Newcomen engine. The steam engine, initially designed to operate pumps for raising water from mines, enabled industries previously reliant on water power to be located in areas remote from streams and rivers. It also enabled improved pumping engines to be built and deeper mines sunk. Trevithick developed a higher-pressure boiler with greater efficiency, smaller cylinders and greater speeds and saw the opportunity to use this to power road and rail transport. He totally changed the breed of steam engine and paved the way for the age of the railway. This led eventually to the two periods of 'railway mania' in 1824-5 and the 1840s when many new railways were proposed and constructed throughout the country.

The Penydarren Locomotive successfully hauled a load of 10 tons on five wagons plus around 70 unofficial passengers, a total of about 25 tons, the 9½ miles from Penydarren ironworks to the canal at Abercynon in just over four hours. Its return journey was delayed until the next day owing to a loss of water after four miles back. It made several other trips but its weight was too much for the cast-iron tramway rails. It was converted to a stationary engine at the works and it was not until 28 years later that a steam engine once again worked the tramway.

However, the principle had been proven and in 1808 Trevithick was asked to build an engine for a mine in Yorkshire. By then he had sold his patent and the next steam locomotive was built by Matthew Murray of Holbeck in 1812 for Middleton Colliery for a three-mile tramway to Leeds. This was the first commercially successful steam locomotive and was followed soon after by others. One incentive apparently was a shortage of horses as a consequence of the demands of the military during the Napoleonic Wars.

14 *A replica of the Penydarren locomotive is in the National Waterfront Museum in Swansea and is steamed on occasional weekends during the year.*

George and Robert Stephenson began building locomotives and supplied a locomotive to Scotts Pit at Llansamlet near Swansea in 1819. This was the second steam locomotive to work in Wales. The Neath Abbey Iron Company had been building stationary engines to Trevithick's design from 1806 and moved into steam locomotive building in 1829. Their first locomotive for a Welsh line was in 1832 when they supplied *Perseverance* to Dowlais Ironworks. Prior to this the Penydarren ironworks had ordered a locomotive from the Stephensons, delivered in 1829. This was used around the works until 1832 when it was returned to them for rebuilding to the wider gauge for the Penydarren Tramroad. The engine, *Eclipse*, returned the Merthyr tramroad to steam power after 28 years.

The development of steam locomotives and the early railways provided much-needed enhanced transport facilities. Many tramroads were converted to railways and new lines built. New railway companies were set up, often by the canal companies themselves or by a similar group of shareholders. The first purpose-designed locomotive-powered line opened in Wales was Brunel's **Taff Vale Railway (S19)** from Cardiff to Abercynon in 1840, extended to Merthyr Tydfil in 1841. Brunel had been commissioned in 1834 while designing the London to Bristol Railway and the Act empowering construction passed in 1836. The TVR ordered two locomotives in August 1840 from Sharp, Roberts & Company which were delivered in September, named *Taff* and *Rhondda*.

The **Llanelly Railway (S39.3)**, opened in 1840 after an empoweringAct in 1836, was another of the early lines in Wales to use steam locomotives, having ordered two from Timothy Hawksworth and Company in 1839, the *Victoria* and the *Albert*. Engineer for the Llanelly Dock and Railway Company and the harbour at Pembrey was George Bush, from about 1835 to 1837 when he moved to be Brunel's resident engineer on the Taff Vale Railway. He had gained his early experience on dock and harbour works under Alexander Nimmo, taking over his work at Aberystwyth after Nimmo's death in 1832.

Other lines followed rapidly and there was a considerable expansion of the railway system over very few years. Railways were proposed for the connection of practically every major town and city in Wales and those over the border in England, several competing for similar routes and many of which failed to reach construction.

Brunel was involved in seven railways in Wales as engineer or advisor, the longest being the **South Wales Railway (S1)**. All except the TVR were built to his broad gauge. The others were the Vale of Neath Railway, the Llynfi Valley Railway, the Ely Valley Railway, the Carmarthen & Cardigan Railway and the South Wales Mineral Railway.

Some of the new railways were built over existing canals. The **Burry Port & Gwendraeth Valley Railway (S38.1)** used the Kidwelly & Llanelly Canal, in places running alongside where space permitted, but using existing canal structures for economy. The Llynfi Valley Railway followed the line of the original Dyffryn Llynfi & Porthcawl Tramroad from Maesteg to Porthcawl, transporting iron and coal to a new dock there.

The new railway companies were promoted in the main by industrialists, ironmasters and colliery owners, who took a major portion of the company shares and as a consequence expected priority treatment for their businesses. This inevitably led to conflict and the promotion of rival railways by the other aggrieved businessmen. An early example of this was the Merthyr Tramroad, described above, promoted by three ironmasters in Merthyr Tydfil because of the priority given to Crawshay of Cyfarthfa Ironworks, a major shareholder in the canal company, of the limited capacity of the Glamorganshire Canal north of Abercynon. Many years later David Davies and fellow industrialists and colliery owners promoted the Barry Dock & Railway Company to provide a new dock and rail link for their coal exports, because of congestion on, and lack of priority treatment by, the Taff Vale Railway and the Bute Docks in Cardiff.

15 *On the breakwater at Porthcawl is a preserved section of the original tramway track.*

The growth of towns and cities at this time, as well as creating a demand for coal, needed building materials. The demand for roofing slate resulted in the rise of the slate quarrying industry in North and Mid Wales. Vast quarry complexes developed and tramroads were built to transport the slate from the quarries to the sea for shipping to the markets. Huge quantities of slate were produced from quarries in Snowdonia, at Nantlle, Bethesda, Llanberis and Blaenau Ffestiniog in the north and Corris and Talyllyn in the south. Most were quarries with large open workings into the mountainside or in deep pits. One at Llechwedd, Blaenau Ffestiniog, however was a deep mine. Slate was discovered in the area around 1775 and full commercial exploitation began in 1799 by William Turner. The slate was mined by forming alternating chambers and pillars carefully aligned through 1,000ft of rock, eventually on 16 different levels forming caverns 50 to 70ft wide and 60 to 80ft high, each taking a team of quarrymen about 15 years to work. The mine is now a tourist attraction with access to the sixth underground level.

Tramways were constructed to convey the slate from these quarries to the coast. The steep-sided and sinuous valleys of Snowdonia meant that these needed to be narrow gauge and when later converted from horse-drawn tramways to railways this narrow gauge was usually retained.

Tramways were built northwards from Penrhyn to Port Penrhyn in 1801 (SH 5973 to SH 6166), from Llanberis to Y Felinheli (Port Dinorwig) in 1824 (SH 5267 to SH 5960), and the **Ffestiniog Railway (N24)** was built westward from Blaenau Ffestiniog to Porthmadog in 1836. The Nantlle slate quarries were connected to Caernarfon Quay in 1828 by the Nantlle Railway. The Stephensons were the contractors for laying the track of this line. It was converted to standard gauge by the Caernarvon Railway and connected to the national network in 1872. It eventually closed in the 1960s.

The Dinorwig Tramroad converted to rail in 1843 as the Padarn Railway, and the Penrhyn Tramroad was replaced by the Penrhyn Railway partly along a different route in 1878. The Ffestiniog Railway has survived virtually intact to the present day again as one of the several narrow gauge railways in North Wales which have been re-opened as tourist attractions. It was closed in 1946 but a preservation society reopened a short length at the western end in 1955 and further lengths later, eventually completing the route to Blaenau Ffestiniog in 1982 where passengers can transfer to the Conwy Valley standard gauge line. The Welsh Highland Railway, a subsidiary of the Ffestiniog Railway, has reinstated a 25-mile length of early 20th-century narrow gauge railways, which closed in 1937, from Caernarfon southwards via Dinas, Rhyd Ddu, and Beddgelert to Porthmadog. A section of the Padarn Railway has been reinstated at Llanberis.

In 1866 the **Talyllyn Railway (N29)** was built to bring slate from the quarries at Bryn Eglws to a link with the Cambrian Railway at Tywyn. It has continued in use to the present day, being taken over by a preservation society when its future was in jeopardy in the 1950s.

In 1902 the **Vale of Rheidol Railway (M2)** was built to convey lead and zinc ore from Devil's Bridge to the harbour at Aberystwyth and is still operating as a tourist attraction. Another narrow-gauge railway was built in Mid Wales in 1903, from Welshpool to Llanfair Caereinion. It closed in 1956 but was reopened by volunteers in 1965 as the **Welshpool and Llanfair Light Railway (M5)**. Other railways were built at Corris in 1859 and Glyn Ceiriog near Chirk in 1873.

From the 1830s the development of railways proceeded apace. In some areas the conversion of horse-drawn tramways was a logical development, in others the replacement of canal systems with a system offering greater flexibility and carrying capacity was the main driver. The engineering of the new form of transport produced significant

16 *Statue of Brunel at Neyland on the site of the old terminus of the South Wales Railway.*

challenges for the civil engineer and has resulted in most of the important historical structures identified elsewhere in this book.

Railways developed westward in both North and South Wales as a means of communication and for the transport of goods. Brunel was asked to plan and build the **South Wales Railway (S1)** which was constructed from Grange Court near Gloucester to Neyland between 1846 and 1856, which he saw as part of his and the GWR's ambitions to link London with Ireland, presenting an opportunity for transatlantic travel.

The main length from Chepstow westward opened in 1850 and the opening of the Wye Bridge at Chepstow in 1852 completed the route to Gloucester. There are many major structures on this route, including his prototype tubular suspension bridge at Chepstow (of which the original foundation piers still remain supporting a later bridge), his river crossings at Loughor (SS 561980) and Carmarthen (SN 405192) and his longest timber viaduct at Landore, Swansea (SS 663959). The rivers at Neath, Loughor and Carmarthen were navigable and the Admiralty required the railway to have opening spans for the passage of shipping with a clear span of 50ft or more. The fixed bridge over the Tawe at Swansea was to have a minimum air draught of 75ft. Over time the navigation requirements were eased and the movable sections have all been replaced or fixed in place.

In North Wales, after rejection of other possible routes such as a possible extension from Shrewsbury via Porthmadog to Porth Dinllaen on the north coast of the Lleyn Peninsula, the **Chester & Holyhead Railway (N9)** was built in 1845-50, with Robert Stephenson as Engineer, to the port of Holyhead, which had been developed as the main port for Ireland more than a generation earlier. This line largely followed the geologically challenging North Wales Coast and involved the construction of a number of important structures such as his new design tubular bridges at Conwy (**N12**) and **Britannia Bridge (N13)** over the Menai Straits, opened in 1848 and 1850 respectively.

In West and Mid Wales other railway enterprises built the Cambrian Coast line and the Central Wales line. Both grew from the amalgamation of various smaller enterprises. The Cambrian Railways arose from the coming together of several railways in 1864, including the Oswestry & Newtown Railway, the Newtown & Machynlleth Railway and the Llanidloes & Newtown Railway, joined in 1865 by the Aberystwyth & Welch Coast Railway. The Central Wales line came from a linking of the Llanelly Railway, the Knighton Railway, the Central Wales Railway and the Central Wales Extension Railway, all of which became part of the London and North Western Railway in 1889. During this period there was considerable interest by, and competition between, railway companies in England to gain access to South Wales. The LNWR gained running powers over lines into Merthyr Tydfil and Cardiff via Hereford and through Mid Wales to Swansea via Llanelli.

17 *Chepstow Tubular Bridge. (© Owen Gibbs collection)*

18 *Loughor viaduct.*

19 *Carmarthen Railway Bridge showing redundant lifting machinery.*

20 *Brunel's original piers at Carmarthen.*

One railway that never came anywhere near the towns after which it was named was the Manchester & Milford Railway. An ambitious project to provide a link from Manchester to Milford for a port for trans-Atlantic trade to rival Liverpool, it opened its only length of track in 1860 from Aberystwyth southward to a junction at Pencader just north of Carmarthen, 40 miles of single track. After abandoning plans (because of the expense) for a route to Aberystwyth via Llanidloes (accessible via the Mid Wales Railway and the Midland Railway to Manchester), it eventually built a line from Aberystwyth to Pencader to use the Carmarthen & Cardigan Railway (which itself never reached Cardigan, only getting as far as Newcastle Emlyn) to Carmarthen and from there used the Pembroke & Tenby Railway to reach its destination at Pembroke on Milford Haven. North of Aberystwyth it used the Cambrian Railways lines to England and on to Manchester. It eventually became part of the GWR in 1911.

As already mentioned, some railways were constructed on the lines of tramroads, using a narrow gauge of about two feet due to the topography of the routes in narrow valleys in North and South Wales. Brunel built his main line, the **South Wales Railway (S1)**, from Gloucester to Neyland at broad gauge but chose the standard gauge of 4ft 8½in for his **Taff Vale Railway (S19)** so as to more easily accommodate the narrowness and curvature of the Taff Valley. He used broad gauge for his South Wales Mineral Railway from Briton Ferry to Glyncorrwg in 1863 as well as his other lines in South Wales. Robert Stephenson chose the standard gauge for his Chester & Holyhead Railway, a gauge that became widely used. Eventually Brunel's seven-foot-gauge lines in England and Wales were converted to the predominant standard gauge.

One of the first combined dock and railway companies in Wales was the Dyffryn Llynfi & Porthcawl Railway, actually a tramroad, in 1828 from Maesteg in the Llynfi Valley to a small harbour and new dock at Porthcawl. The line was converted to a railway in 1861. The port exported coal and iron products and imported zinc ore for smelting in the valley.

Further west is one of the oldest man-made harbours on the South Wales Coast and another of the oldest combined dock and railway companies.

The first Pembrey Harbour was opened in 1819 by the Pembrey Harbour Company to export coal from the Gwendraeth Valley along a new canal. It had siltation problems and in 1825 a new company, the Pembrey Harbour & Railway Company, was formed and built the New Harbour half a mile eastwards, which was opened in 1832 and became known as Burry Port Harbour.

21 *Entrance to the Harbour at Porthcawl.*

Wales has many firsts and one is the first passenger-carrying railway in the world and the first railway timetable. In 1807 the Oystermouth Railway opened for passenger traffic along Swansea Bay. The line was built in 1806 as a tramway to a four-foot gauge to convey coal and limestone to the Swansea Canal from Oystermouth by horse-drawn drams. It was extended to Mumbles Pier in 1896, being known as the **Swansea and Mumbles Railway (S35)**. The carriages were converted to steam hauled in 1877 and the line electrified in 1929 using double-decker cars and a pantograph system. The line was closed in 1960 and part of the route is now a footway and cyclepath along the foreshore. The station at Blackpill is still in use as a café. (SS 620905).

Another unusual railway in Wales is the **Snowdon Mountain Railway (N28)**. This was built between 1894 and 1896 as a tourist venture and is the only rack and pinion railway in Wales. It was fully opened in 1897. It runs for over four miles from Llanberis to the peak of Snowdon, the highest railway station in Britain at 3,560ft, from where (on a fine day) there are wonderful views over Snowdonia. Initially proposed 20 years earlier it was not until the beginnings of the decline of the slate industry that the landowner accepted the need to encourage visitors to the area as an additional business. It carried about 12,000 passengers in its first year of full operation and now carries around 140,000. The line is too steep for a conventional railway, rising over 3,000ft, hence the need for the rack and pinion system to provide a positive connection between the engine and the rails. It was not, however, the first rack and pinion railway. This distinction goes to the Middleton Colliery Railway, which ran three miles to Leeds, built in 1812. Here on a relatively level track a third toothed rail and cogged gear on the locomotive was used

22 *Blackpill Station on the Oystermouth Railway. (© Owen Gibbs collection)*

to improve traction and permit a lighter engine to be used, overcoming Trevithick's problem at the time of broken rails with heavier locomotives.

At Aberystwyth there is a cliff railway (**M1**) 778ft long, which is a cable-drawn balanced system of two passenger cars on a continuous wire rope system. It rises from the promenade to a view point on the headland and is the longest in Britain. The Great Orme Tramway in Llandudno (SH 778828), opened in 1902, is a cable-hauled system and Britain's only remaining example. It rises about 550ft over a mile or so. The winding house is situated at the halfway point.

ROADS

In prehistoric times roads were merely tracks, usually along ridgelines for ease of travel, avoiding dense forests and swampy ground. Spurs down valleys to fertile areas were made along streams and rivers. In Wales the Romans created major paved roads over most of the country, initially for military use, later for access to minerals such as lead, silver and copper in North Wales, lead and silver in Mid Wales and gold in West Wales. Roman roads were usually built in straight sections between forts at river crossings, which were usually fords, or followed the major rivers inland. Roads were built across South Wales from Caerleon to Carmarthen and across North Wales from Chester to Caernarfon and Anglesey. A south-north route was built from Carmarthen to Trawsfynydd (Tomen-y-Mur) and on to Caernarfon and roads through Mid Wales linking Caerleon, Cardiff and Carmarthen via Brecon to Wroxeter (near Shrewsbury). Recently a broad Roman road was discovered during the construction of a new bypass at Whitland west of Carmarthen leading towards Milford Haven.

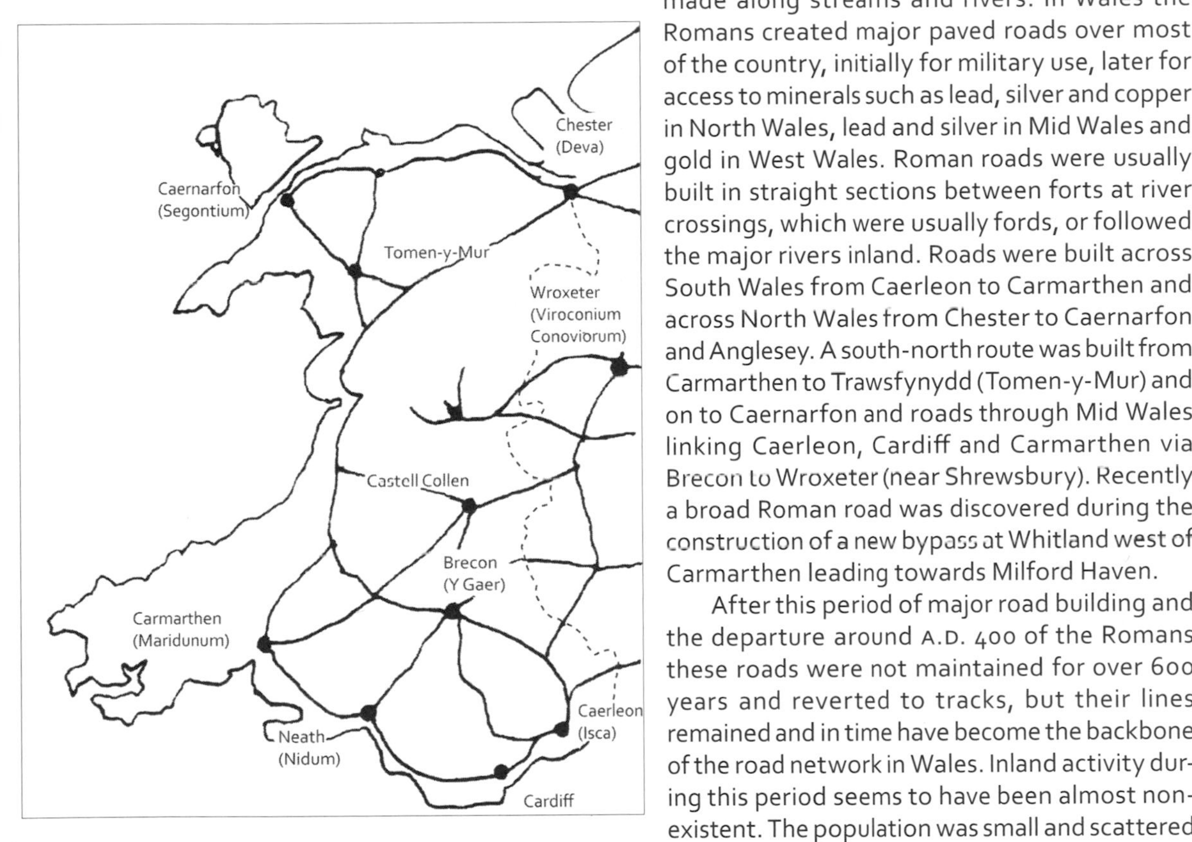

23 *Roman roads in Wales.*

After this period of major road building and the departure around A.D. 400 of the Romans these roads were not maintained for over 600 years and reverted to tracks, but their lines remained and in time have become the backbone of the road network in Wales. Inland activity during this period seems to have been almost non-existent. The population was small and scattered with few settlements larger than villages. What bridges there were, usually timber ones left from the earlier Roman period, were not maintained unless of strategic value and river crossings were by ford or boat.

In the early medieval period rough tracks and narrow bridleways were maintained by religious orders and by Boroughs by Royal Charter. Transport at this time would have been mainly by packhorse and coastal shipping. Towns were generally self-sufficient and there was little need for long-distance travel. In Tudor times after dissolution of the monasteries, who were one of the principal agencies for maintaining the roads, the responsibility for road maintenance was placed upon the parishes by an Act in 1555, to be under the control of a local parish surveyor (unpaid and unqualified!), and local rates levied for the work. The Act required the parishioners to work for several days a year on road maintenance or pay a rate in lieu. This was the first time there was a statutory responsibility for highways. Naturally the parish was only interested in their local needs and there was no provision for long-distance travel. Bridges, however, were administered on a county basis by Justices of the Peace at Quarter Sessions, who were given powers to repair bridges and levy rates and fines for so doing.

In the early 18th century the growth in the export of cattle to England led to the expansion of drovers' roads. Roads at this time were merely tracks, described as full of dust in summer and virtually impassable in winter. One drovers' road grew in importance from West and Mid Wales to England via the Usk valley, dividing at Brecon southward down the valley to Monmouth and eastward via the Llynfi and Wye to Hereford.

The demand for improved communications and the development of roads from poor quality, under-maintained, lanes and drovers' roads controlled by the local parishes led in the late 17th and 18th centuries to Parliamentary authority for the formation of Turnpike Trusts. These were given powers to borrow money to improve and maintain the roads in return for levying tolls. Eventually over 23,000 miles of road and 1,100 Turnpike Trusts existed. In Wales there were five turnpikes in the period 1663-1700, a further 144 by 1750 and 316 by 1779. Mail coaches were introduced around 1783 and Government Mail coaches in 1784. In 1787 the Irish mail used the South Wales route for the first time. In 1788 a regular coach service ran three times a week from Swansea to Bristol via the New Passage Ferry from Portskewett in Monmouthshire to Redwick in England. By 1830 there were 12,000 miles of coaching routes in Britain. For 50 years these provided the main form of long-distance travel until rail travel took over. The last mail coach ran in 1850.

From the mid-18th century the techniques of road construction were being developed beyond simple filling of potholes and damaged verges, with greater thought to the road construction itself to cater for the increased traffic. In the early 19th century Telford and McAdam developed independently a layered construction of a base layer with side and cross drains to provide effective drainage of the road base, a support layer to distribute the load and reduce the pressure on the original surface and a wearing layer of finer material to produce a good running surface. In towns the use of stone paving and stone setts was common and also in places wood blocks to provide a more durable surface.

Not unsurprisingly, turnpikes were not universally welcomed and between 1839 and 1844 the 'Daughters of Rebecca' in West Wales burnt many of the new toll gates in night-time attacks. This was a symptom of the poverty, severe economic depression and general unrest at the time, not only in Wales but throughout the kingdom. The activities spread to attacks on the landowners and tithe collectors. Eventually it led to military intervention. As a consequence in 1845 a County Roads Board for South Wales was set up by Act of Parliament to take control of the turnpikes and main roads.

In 1835 the Highways Act marked the beginning of modern highways administration, retaining the parish duties but authorising the appointment of highway surveyors and the levying of highway rates on local landowners. The 1862 Highways Act combined most parishes for road administration purposes with the office of a county surveyor. The Highways and Locomotives Amendment Act of 1878 abolished turnpikes and set up County Highway Authorities. Turnpikes had been declining anyway with the development of railways. The last turnpike trust ended in 1885 and the last tollgate on the London to Holyhead road in 1895.

The formation of County Highway Authorities brought new organisation and ideas to road construction and the development of the motor vehicle introduced new demands for good quality highways. Counties and county boroughs assumed responsibility for

24 *Welsh Major Roads, 2009.*

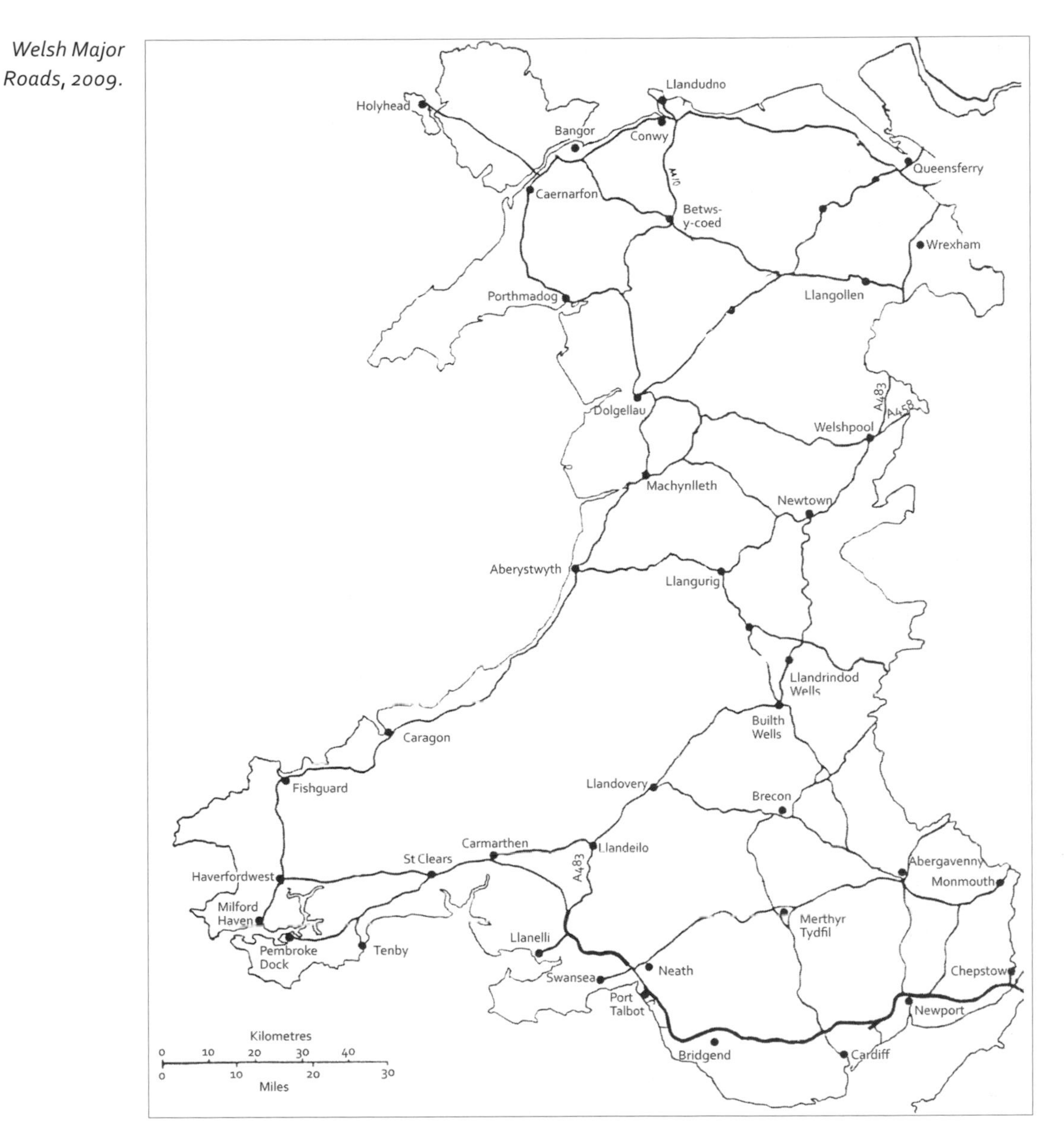

roads and bridges in their area and their Highway Departments became responsible for maintenance and improvement as traffic increased in volume, size and weight. The Local Government Act of 1888 had introduced county councils and county boroughs and included the repair of roads and bridges in their responsibilities. Urban and Rural Districts were established in 1894 and some counties delegated their highway responsibilities.

However, little was done to improve the road network. Motor vehicle numbers began to increase and in 1910 a Roads Board was set up to give grants from central government to local authorities for road maintenance. Its main achievement in its short life to 1919 when the Ministry of Transport was established was the sealing of carriageways with tar or bitumen. From the turn of the 19th century the motor vehicle with its greater speeds damaged the water-bound road surface with a consequent rapid deterioration. Thus began the sealing of the surface with tar, a by-product of the gas industry, and later bitumen, a by-product of the petroleum industry, to bind the surface material.

In the 1920s a special programme of road improvements was introduced to provide work for the unemployed. In Wales the Glamorgan Valley Roads Project provided a series of links between adjacent valleys in west Glamorganshire. Five links were built between several separate valleys over difficult inter-valley terrain.

In 1936 30 major routes, 4,460 miles, in Great Britain were designated Trunk Roads and taken over by the Ministry of Transport, who became responsible for the construction and maintenance of these strategic routes. This was increased to 8,232 miles in 1946. The need for a strategic network had become apparent, much as had the government-sponsored London to Holyhead Road in the 19th century. Different priorities by adjacent counties for highway improvement meant a lack of continuity of standards on major cross-country routes and a national overview was necessary to plan the route development and maintenance on a country rather than county basis. The development of the strategic motorway network by the Ministry, authorised by the Special Roads Act of 1949, commenced with the opening of the first length in 1958.

Four years after the formation of the Welsh Office in 1965 trunk roads and the motorway in Wales, previously managed by the Ministry of Transport, became a Welsh responsibility. The Welsh Office was replaced by the National Assembly for Wales and the Welsh Assembly Government in 1999.

Telford's Road Projects in the 19th Century

Several major route improvements were engineered in Wales. The most significant of these was the development by Thomas Telford of the **London to Holyhead Road (N1)** now the A5. In 1810 Telford was commissioned by Parliament to survey the route for improvement. Work began in 1815 with Telford concentrating on the worst sections of the road in Wales. By 1819 he was able to report that the worst sections had been improved. At this time the **Menai Suspension Bridge (N4)**, was still unfinished and the crossing of the Menai Strait was still by the existing ferry from Bangor. Menai Bridge was completed in 1826. The A5 was one of the first government-sponsored highways and was needed to speed the link to Ireland. It reduced the journey time from Shrewsbury to Holyhead from 17 hours to about twelve. Telford engineered a new form of construction to suit the increasing traffic with better materials and properly designed maintenance to ensure his route could be travelled at speed. Gradients were limited to maintain an average speed of 10mph wherever possible and the whole route maintained to a high standard with frequent coaching inns for changes of horses as part of the project.

In 2003 Cadw published a report of a survey of the A5 in Wales identifying extant features of Telford's work. On much of the alignment there is still considerable original work to be seen. Cadw has estimated that over the whole route from Chirk to Holyhead at least 35 per cent can still be seen and in some sections as much as 65 per cent is still intact. The items include original walling, maintenance depots (small laybys every quarter of a mile which held stocks of road repair material), earthworks and retaining walls, toll houses, several of his 'sunburst' pattern gates, and of course many of his bridges. Menai Suspension Bridge and **Waterloo Bridge (N2)**, Betws-y-Coed are probably the best known. Examples of his retaining walls can be seen at Glyn Diffwys Bends (SH 995443 to SH 991444), where his original road has been kept as a footpath when the section was bypassed in 1998, and the **Nant Ffrancon Pass (N3)** through Snowdonia.

On the mainland length of the Holyhead road Telford used a standard design single-storey tollhouse. For the tollhouses on Anglesey he adopted a two-storey design as at Llanfair P.G. (**N1.1**). It is not known why he used the two different types. The best single storey example is at Ty Isaf (**N1.1**) four miles west of Llangollen and one has been re-erected at Blists Hill Museum, Ironbridge. Alongside his suspension bridge at Conwy

25 *Telford's historic road. (© PB)*

(**N8**) on the route between Chester and Bangor he built a very ornate mock-medieval tollhouse to a similar plan as his standard single-storey tollhouse which he used elsewhere.

In March 1998 the Welsh Office designated Telford's Holyhead Road across Wales, the A5 Trunk Road, an Historic Route and erected information signs to this effect at each end with repeater signs at intervals. The intention is to retain as many of the original features of the route as possible with only very selective improvements for road safety.

Cadw's survey had identified that many of Telford's original milestones were still in place, some with their original cast-iron plates. As a continuation of the Welsh Office policy to designate the route an historic road, in 2003 Gwynedd Council, on behalf of the new Welsh Assembly Government, refurbished the existing milestones and added five new ones where they were missing, using the same stone as the originals from Moelfre, Anglesey. Additionally, 39 new cast-iron plates were made to the original pattern and fixed to recreate the 84 milestones originally on the route.

Telford was also commissioned to rebuild and renew the route between Chester and Bangor through Conwy. This was to follow generally the already existing route along the coast established by earlier turnpike trusts. The Cadw survey of 2003 found little evidence of his work now surviving. The main section still to be seen was the **Conwy Suspension Bridge (N8)** and his route further west over the headlands each side of Penmaenmawr. This section was bypassed in the 1930s by a new route around the headlands, partly in tunnel and in places on viaduct on the cliff edge.

Prior to the completion of the bridge at Conwy the only bridge over the river was several miles inland at Tal y Cafn, where a route led over the high ground towards Bangor on the line of the old Roman Road. There was also a ferry across the River Conwy with several tracks across the headlands at Penmaenmawr and Pen y Clip which were passable with care and always the alternative option around the headlands across the sands at low tide. Telford improved the headland route as part of the work and this part of Telford's road has now been adapted as a cycle route to avoid the new tunnels built in the 1990s as part of the upgraded A55. A booklet written in 1935 by H.L. North at the time of the upgrading of the headland road at Penmaenmawr is entitled *The Seven Roads across Penmaenmawr.* He lists these as an ancient trackway (date unknown); the First Lower Road (*c.*1700-20), suitable for coaches (with difficulty) as an alternative to the route across the sands; Sylvester's Road (1772), which improved the lower road; Telford's Road (1830); later alterations (1840-60); Stephenson's Railway (1845); and the New Road (1935). To this can now be added the new A55.

26 *The original toll board at Llanfairpwllgwyngyll Anglesey.*

Tolls to be taken at
LLANFAIR GATE
s. d.
For every Horse, Mule, or other Cattle drawing any Coach or other Carriage with springs the sum of 4
For every Horse, mule or other Beast or Cattle drawing any Waggon, Cart, or other such Carriage, not employed solely in carrying or going empty to fetch Lime for manure the sum of 3
For every Horse, Mule or other Beast or Cattle drawing any Waggon, Cart, or other such Carriage, employed solely in carrying or going empty to fetch Lime for manure the sum of 1½
For every Horse, Mule or Ass, laden or unladen, and not drawing, the sum of 1
For every Drove of Oxen, Cows, or other neat Cattle per score, the sum of 10
For every Drove of Calves, Sheep, Lambs or Pigs per score, the sum of 5
For every Horse, Mule or other Beast drawing any Waggon, or Cart the Wheels being less than 3 inches in breadth, or having Wheels with Tires fastened with Nails projecting and not countersunk to pay double Toll.
A Ticket taken here clears Carnedd Du Bar.

Telford's bridge at Conwy remained in use until 1958 when a new single carriageway steel arch bridge (**N17**) was built alongside. As well as relieving the bridge of the increasing traffic volumes and weights this was intended eventually to be part of a dual carriageway bypass of the town which would go along the quayside and around the north of the town. It would have involved the eventual demolition of Telford's Suspension Bridge. The proposal met with considerable opposition and the proposed second carriageway was eventually abandoned, which explains the apparently unfinished condition of the north elevation.

In the 1970s the upgrading of the whole route from Chester to Holyhead to a dual two-lane high-speed highway was planned by the Welsh Office and the method of bypassing the narrow congested streets of Conwy was reconsidered as part of the route study. This initially proposed a new route to the north with a new bridge further downstream from Deganwy to Conwy Morfa. Following objections the Welsh Office submitted an alternative for a bridge upstream of the town with a tunnel under Conwy Mountain. In 1980 the Secretary of State, after the longest public enquiry in highway history at the time, announced his decision to build a more expensive tunnel under the estuary to the north of the town, on the line of the proposed bridge, to safeguard the views of this World Heritage Site. **Conwy Tunnel (N18)** was the first immersed tube tunnel constructed in the UK and was opened to traffic in 1991. An immersed tube tunnel is formed by building large tunnel sections nearby, floating them out and sinking them into a deep trench excavated in the river bed. After sinking and connecting the units they are then covered with material and the river bed re-instated.

In 1826 Telford also reported on route options across South Wales. These included one from Bristol via a ferry across the River Severn at New Passage to Monmouthshire and then via Cardiff and Swansea to Pembroke (and steam packet to Waterford in Ireland) as an option to the then existing route via Gloucester and Chepstow to Cardiff, Swansea and Carmarthen (now the A48). However, the Holyhead route proved more suitable with a direct link to Dublin. The other route across South Wales considered was the London to Fishguard Road via Gloucester, now the A40 trunk road, which enters Wales at Monmouth and follows the Usk Valley to Brecon and then the Towy Valley to Carmarthen and beyond, much as the medieval tracks did. There is no evidence of the earlier construction other than perhaps the medieval bridge at Monmouth (**S5**) and the alignment of the old road through the various towns and villages along it.

Both these routes were in use in medieval times from England to Pembroke for transit to Ireland and are protected by Norman castles at key points. In addition, in the 13th century Henry I settled Flemish immigrants in South Pembrokeshire to anglicise the region as a further safeguard to the route to Pembroke and Milford.

27 *Carew Castle, Pembrokeshire.*

BRIDGES, VIADUCTS AND AQUEDUCTS

A bridge is generally accepted as crossing a single feature such as a river whereas a viaduct crosses several such as in a valley. An aqueduct carries water over either. All can be single or multiple spans.

BRIDGES

Bridge construction has been a major feature of the work of the civil engineer from time immemorial. Design and construction has responded to changes in form, size and materials from simple beams in timber through arches in stone or brick and cast iron, trusses in iron and steel, to concrete in many forms and large span steel girders and suspension bridges.

Some are major ones such as Telford's **Menai Bridge (N4)**. Others may seem on initial view to be insignificant but often represent the early development of new techniques and materials, for example the cast-iron **Pont-y-Cafnau Bridge (S24)** of 1793 at Merthyr Tydfil, probably the world's earliest cast-iron tramroad bridge. Many medieval bridges still exist, albeit often somewhat altered over the years. Early river crossings were by ford or ferry and the location of villages and towns often indicate the original crossing place. From Roman times some permanent structures were constructed for highway use. Later, as canals and tramways were built, followed by railways, new forms of construction were developed using new materials and with greater and greater spans.

Bridges have been categorised by the Institution's Panel for Historical Engineering Works into masonry, cast iron, wrought iron and concrete construction for convenience and have been assessed for relative importance in engineering history using eight criteria. These include age, innovation, size, rarity, aesthetics and status of the designer and contractor. This has enabled a list of the most significant to be identified. Many bridges and viaducts in Wales that have been listed by Cadw do not appear in the Panel's list because of these very selective criteria.

There are thought to be about 75,000 bridges in Great Britain. The earliest structures would have been quite simple. The clapper bridges on Dartmoor probably represent one tradition of bridge building from the earliest times. The Romans seem to have used timber superstructures on timber or stone abutments and piers, although some major bridges had stone arches. Four hundred years later from the eighth century the building of new bridges began again. Their existence is known mainly from references in charters. Most were of timber which required regular maintenance and replacement. Medieval records frequently show that many were in poor condition and in need of repair. Grants of 'pontage' were made at the Assizes or Sessions, usually for a three-year period, to allow tolls or taxes to be levied to fund repairs and reconstruction. The Court of Quarter Sessions would decide if a bridge was out of repair and who should be responsible for its maintenance, levying a fine if no action had been taken to carry out the work. Only if no other body could be found to maintain the bridge would the justices take the responsibility. Bridges were sometimes built by religious orders and often would include a small chapel with a priest to collect alms for the maintenance and say prayers for the souls of the donors.

Over time timber bridges would be replaced with more substantial masonry ones. From the 11th century stone arch masonry bridges were built. In the 15th century some were built of brick although timber continued to be used. No timber highway bridges survive from these times and only about 200 masonry ones retain all or part of their medieval fabric. However, there are timber bridges at many of the medieval castles in

Wales; these have been rebuilt several times but generally retain the original form in keeping with the ancient buildings.

The designers of the early tramways and railways also used timber for bridges and viaducts. In Wales Brunel used timber for many of his viaducts including Landore (SS 663959) and Loughor (SS 561980), the latter still in use although Brunel's timber bays have been reconstructed above low-tide level and the deck replaced. Two of his timber viaducts were still in use until the 1940s at Aberdare, being demolished in 1947 after the line closed.

One of his timber viaducts, 1,200ft long with a 100ft navigational span, at Newport on the South Wales Railway burnt down in 1848 just before the line was ready to open and was hastily rebuilt with a new centre span, a bowstring girder, in wrought iron which could be fabricated while the rest of the timber was replaced, speeding the rebuild.

Other timber viaducts still exist on the Cambrian Coast railway. The longest is at Barmouth (**N31**) although some of the timberwork will have been renewed over time. In the 1980s major repairs were required because of an attack by a marine borer, teredo worm, and locomotive-hauled trains were prohibited for some time.

From the late 18th century cast iron was used, first in arch form and later for beams, gradually replaced in the 19th century by wrought iron, particularly for trussed spans. Later in that century wrought iron was replaced by steel and from the early years of the 20th century reinforced concrete became preferred to masonry for arch structures. Prestressed concrete made its appearance during the austerity years at the end of the Second World War and composite steel/reinforced concrete bridges about twenty years later. For longer spans steel box-girder bridges were developed.

The earliest comprehensive compilation of Welsh bridges was by Edward Lhwyd in the early 18th century. Edward Lhwyd (or Lluyd) in 1690 became keeper of the Ashmolean Museum in Oxford where he had studied. He had a wide range of interests, geological, biological and antiquarian, and undertook a series of tours around Britain, including Wales between 1697 and 1701. His information on Welsh bridges was used by Edwin Jervoise in his book on ancient bridges in Wales published in 1936. He identified most of the bridges in Lhwyd's works and found 70 per cent of them. Ogilby's Road Maps of 1675 also show the various bridges along his routes which are a further source of information as are John Leland's accounts of his travels in Wales in the early 16th century.

Masonry Bridges

Ascribing dates to ancient bridges is difficult because of the lack of contemporary records. Even where accounts do exist, the present structure may be a subsequent rebuilding that has passed unrecorded. There were no clear transitions from one style to another but some clues may be found in the arch shape. The earliest arches were either semi-circular, in the Romanesque tradition, or curved-pointed Gothic arches. From the 14th century segmental arches enabled the construction of flatter shapes and therefore less steep approaches, important when the motive power was provided by horses. Gothic arches became flatter, too, and in the 16th century became four-centred, perpendicular, or Tudor in shape. Elliptical arches were introduced to Britain in the 1760s. Many medieval bridges were generally narrow, suitable for packhorses only. These were later widened to take carts and wagons. Sometimes it is possible to identify the later work on one or both sides of the bridge.

Increased commercial activity and the spread of turnpike roads from the early 18th century led to an extensive programme of upgrading and new construction, with the traditional designer/masons replaced by architects and engineers and a new age of elegance. Almost half of the masonry bridges in Britain were canal and railway over-

28, 29 *Bridgend Old Town Bridge.*

and under-bridges. Standard bridge designs that could be adapted by the resident engineers to suit the individual locations were produced by designers based away from the construction sites. There was, however, still scope for virtuosity.

Of the medieval and post-medieval bridges in Wales the oldest ones still remaining are **Monnow Bridge (S5)** at Monmouth (13th-century), the only remaining fortified bridge in Britain; **Devil's Bridge (M3)** in the Rheidol Valley (said to be 11th-century and known from records to exist in the 14th); **Llangollen Ancient Bridge (N36)** (late 15th-century); Old Town Bridge at Bridgend (15th-century) (SS 904798); **Pont Spwdwr (S40)** near Kidwelly (16th-century); Pont Abercamlais (1580) (SN 965291), and **New Inn Bridge (S30)** (pre-17th-century).

Seventeenth-century masonry bridges include **Pont Fawr Llanrwst (N16)** (1636), considered to be one of the most attractive in Wales; Bangor Iscoed (*c*.1650) (SJ 389456); Llechryd Bridge over the Teifi (1656) (SN 218436), notable for its very low parapets to allow for the flooding of the river; **Cysyllte Ancient Bridge (N39.1)** (1697); Leckwith near Cardiff (17th-century) (ST 159752); the five-arch **Pont Carrog (N35.1)** (1661); **Pont Cilan (N35.4)** (pre-1700); and **Llangynider Bridge (S8)** (1700).

Other bridges of similar antiquity are the seven-arch bridge at Corwen (**N35.2**), the longest on the Dee (1704); **Crickhowell Bridge (S7)** (1706); **Pont Dyfrydwy (N35.3)** (*c*.1700); and Aberffraw (1731) (SH 355689). Notable early 19th-century masonry bridges include **Pontygwaith Bridge (S23)** (1811) and the unusual design three-arch **Pant y Goytre Bridge (S6)** (1826).

William Edwards is the best known of the 18th-century bridge builders in South Wales. His bridge at Pont-y-tŷ-pridd (**S21**) (now Pontypridd), said to be at 140ft the biggest masonry arch span in Britain when built in 1756, and Dolauhirion (**M16**) built by him in 1773, remain today as fine examples of his work. The bridge at Pontypridd was where he gained much experience in the science of masonry bridges.

Born in 1719 at Tŷ Canol, Groeswen, near Caerphilly in Glamorganshire he and his family moved to a farm at nearby Bryn Tail where he lived until his death in 1789. He became an Independent minister and was pastor at Groeswen Independent Chapel from 1745, remaining so all his life. He had been interested in masonry and building from an early age. The youngest son of the family, he was just seven when his father drowned while fording the River Taff. By the age of 15 he was repairing stone walls for neighbouring farms to help the finances and became interested in stone-masonry after seeing masons

30 *Pontardawe Bridge. (© Owen Gibbs Collection)*

building a smithy nearby. At the age of 20 he constructed several buildings including a mill at Craig-y-Fedw near Abertridwr.

In 1752 he was contracted to build a bridge over the River Taff at the tiny hamlet of Pont-y-tŷ-pridd for the sum of £500, one that would stand for seven years. It took four attempts. The first, a three-span arch bridge, stood for just over two years before being demolished by debris brought down in a flood. He then built a single span arch across the river but the temporary supports were washed away before completion. The third attempt followed with more substantial formwork but on striking the supports he saw the bridge collapse, the centre moving upwards due to the pressure from the substantial abutments. His fourth attempt in 1755 was successful after he redesigned the haunches with large circular voids to reduce the weight and increased it at the crown. This bridge still stands and can be seen alongside a new bridge opened in 1857.

From the experience gained from this project he and his sons built many bridges across South Wales, notably Dolauhirion (1773), still in use as a highway bridge, and others

31 *Wychtree Bridge, Morriston, c.1950. (© Owen Gibbs Collection)*

32 *An illustration of Newport Bridge from the north by Sir R.C. Hoare in* An Historical Tour in Monmouthshire *by William Coxe, 1801. (© Newportpast)*

at Usk (*c.*1760), where his five-span bridge, enlarged in 1836, is still in use; Pontardawe (1757), an 80ft arch of which little is visible after several widenings; Beaufort Bridge over the Tawe at Llangyfelach just north of Swansea (a three-arch bridge contemporary with his first bridge at Pontypridd), removed in 1968; Wychtree Bridge at Morriston (*c.*1780), a 95ft single arch, demolished in 1959 to make way for a new highway bridge; a 70ft arch over the River Afan at Aberavon (1768); a 45ft-span single arch over the River Aman at Betws, Rhydaman (*c.*1770); and a five-arch bridge at Glasbury. This bridge of 1777 over the Wye was carried away in a flood of water and ice in 1795 and replaced in timber.

He may also have built the bridge at Tredunnock, another three-arch bridge just south of Usk, and a bridge at Pontycymmer. In addition to all this he was employed by (Sir) John Morris, owner of Llangyfelach (Beaufort) and Forest copper works, to plan his new village of Morriston, near Swansea, in 1768 and to superintend its construction over the next 20 years. It is also believed that in the mid-1760s, after working at Cyfarthfa Ironworks, he was involved in the building of several bridges on Kymer's Canal,

33 *Cenarth Bridge over the River Teifi.*

including Pwll-y-Llygod tramroad bridge over the Gwendraeth Fawr, possibly the first in Wales. Although somewhat covered in vegetation this Scheduled Ancient Monument can still be seen (SN 444067).

He was much in demand for his engineering expertise in bridge building and repair work. He and his sons and their masons were fully employed. Between them they built several other bridges, including Edwinsford over the River Cothi at Talley, Carmarthenshire (1783) which still stands and a bridge over the River Rhymney at Bedwas.

His son David became leader of the family business with his brothers Thomas and William after their father's death in 1789. A bridge at Llandeilo ascribed to David Edwards is possibly the bridge at Llandilo yr Ynys, Nantgaredig (1786) (SN 492204), a few miles downriver. Newport Bridge, another five-arch bridge, was built in 1801; it was replaced in 1925. David Edwards was also the designer of the three-arch Cenarth Bridge (1788) (SN 269415), although not the builder.

34 *Ynysgau Bridge c.1936. (© Society for Protection of Ancient Buildings)*

Cast-Iron Bridges

The early cast-iron bridges emulated the construction forms of timber or masonry, but using the new material. Only 14 bridges and aqueducts survive from the first 30 years, up to 1810, in the British Isles, including two in Wales, **Pontcysyllte Aqueduct (N39)** and **Pont-y-Cafnau Bridge (S24)**.

In Wales the earliest cast-iron bridge is the tramroad bridge Pont-y-Cafnau, mentioned above, built in 1793 in Merthyr Tydfil. This probably replaced an earlier timber bridge to the Cyfarthfa Ironworks. It carried an aqueduct at high level and another at low level with a tramway. An aqueduct over 600ft long and 80ft above river level, also in iron, brought water from the river to power a large water wheel at the ironworks. The cast-iron joints are based on the timber jointing method in use at the time.

Another of the early cast-iron bridges in Wales was Ynysgau Bridge in Merthyr built in 1800. It was designed by Watkin George, engineer at Crawshay's Cyfarthfa Ironworks, and paid for by William Crawshay to replace a stone bridge that had collapsed in 1795 after a severe flood. With only one other bridge across the river in the area Crawshay needed a crossing to his Cyfarthfa Works from the main town. Work started in 1799. George realised that to bridge the Taff in a single span of about 70ft was beyond the capacity of single cast-iron beams and designed the bridge as a series of cast-iron beams and frames in three sections in a shallow arch. It remained in full use until about 1880 when a larger bridge was built nearby and then remained in use as a footbridge. It was eventually demolished in 1963 but most of the metalwork is still stored in Cyfarthfa Park, Merthyr, awaiting funding for re-erection. **Robertstown Tramroad Bridge (S27)** of 1811 over Aberdare Canal is another one of the world's oldest surviving cast-iron bridges.

From 1810 a number of designs emerged that used cast-iron to its best effect, in various configurations of arches. Thomas Telford produced an elegant standard cast-iron arch design which he used for his bridges, with spans of 105ft and 150ft (and one of 170ft). The bridge by him at **Betws y Coed (N2)** on the Holyhead Road, cast in 1815 by William Hazledine and erected in 1816, is a fine example.

Another early cast-iron bridge in Wales is **Chepstow Bridge (S2)**, designed by John Rastrick and cast by Richard Hazledine of Bridgnorth (1816). Richard was the brother of William but independent of him; the firm was Hazledine & Rastrick. Other early cast-iron bridges include Aber Ogwen (1824) (SH 611722) near Bangor, cast at Penydarren Works; **Bigsweir (S4)**, 164ft span (1828); **Llandinam (M7)**, a 90ft-span bridge by Hawarden Ironfounders (1846) using patterns from Hazledine's Kynaston Foundry;

Abermule (M8), a 110ft-span bridge by Brymbo Ironworks (1853); and **Caerhywel (M9)**, two 72ft spans by Brymbo Ironworks (1858). Brymbo Ironworks was founded in 1793 by John Wilkinson, who had established Bersham Ironworks in Wrexham in 1762. These last three bridges were designed by Thomas Penson junior, County Surveyor of Montgomeryshire and Denbighshire. He had produced a cast-iron design in association with Telford in 1824 for Llanymynech (SJ 267205) which was put aside in favour of an elegant masonry bridge instead.

Wrought-Iron Bridges

Cast iron is relatively weak in tension and therefore not ideal for use as beams. Robert Stephenson experienced problems with cast iron as late as 1847 when part of his Dee Bridge on the Chester & Holyhead Railway collapsed, leading to one of the Railway Inspectorate's earliest fatal accident enquiries and a review of the use of cast-iron beams. The manufacturing process was not entirely reliable, giving rise occasionally to unexpected brittleness or voids in the castings, and although longer lengths were sometimes used most engineers limited themselves to 30ft spans.

Wrought iron offered many of the advantages of cast iron with the additional one of being strong in tension as well as compression. Until James Neilson patented his hot blast method of manufacture in 1828, the cost of the product prevented its use for bridgework, but in 1831 a girder bridge of moderate span was built over a railway in Glasgow. Wrought-iron flat links up to nine feet long from Hazledine's Upton Magma foundry were used for both of Telford's Welsh suspension bridges. At Menai Bridge about 15,500 were used together with other wrought-iron elements. Robert Stephenson's innovative record-breaking tubular bridges at Menai and Conwy (**N12** and **N13**) in 1848-50 on the Chester & Holyhead Railway were constructed in wrought iron using shipbuilding techniques following considerable testing by shipbuilder William Fairburn. The process of boring holes for the rivets, over two million at Britannia Bridge alone, in the plates was greatly accelerated and improved by the development by

35 *Crumlin Viaduct. (© Owen Gibbs Collection)*

Richard Roberts of Llanymynech of a precision punching machine.

Of greater significance was the development of a number of truss types that could use wrought iron to provide economical spans up to 250ft. **Crumlin Viaduct (S16)** was built in 1857 with wrought-iron Warren trusses for the beams and columns. The columns were 200ft high, the maximum span of the beams was 150ft. The viaduct was demolished in 1966 but the massive abutments can still be seen each side of the valley (ST 213986).

36 *Briton Ferry Viaduct.*

Steel Bridges

Steel replaced wrought iron for some uses from the 1850s, but was not used in bridges until the 1880s. The Forth Bridge, opened in 1890, was built of steel, and thereafter wrought iron was almost entirely superseded. Welsh steel bridges include **Hawarden Railway Swing Bridge (N32)**, which was the longest movable span in the world when built in 1889. Carmarthen Railway Bridge (SN 405192) was built in 1911 to replace an earlier Brunel bridge of 1854 on the South Wales Railway. Brunel's foundation piers still remain. The bridge had a Scherzer-type rolling lift section to allow shipping to pass upriver; this is now fixed in place. Another late railway swing bridge, thought to be the only one on a curve and a skew, was built over the River Neath in 1894 by the Rhondda & Swansea Bay Railway (SS 730964). It was incorporated into a Swansea Avoiding Line for goods traffic by the GWR. The swinging span is now fixed.

Newport Transporter Bridge (S14) was built in 1906 and is one of the few remaining bridges of its type in the world. It has steel lattice girders supported on steel lattice towers to carry a moving platform at road level across the river as an aerial ferry. It was built to provide access to the east side of the river for industrial development.

Queensferry Bridge (SJ 322687) is an early steel truss bascule lift bridge with steel plate cross-girders built in 1926. It has two similar Scherzer rolling leaves. There was a Scherzer bridge in Swansea Docks which has now been dismantled. William Scherzer developed the rolling bascule bridge in America in 1893. It had advantages over the standard bascule bridge, being lighter and allowing the bridge to span greater distances as well as providing greater clearance over the waterway when it was rolled back. Briton Ferry Viaduct (SS 730943) is a steel-plate girder with concrete deck, the first large post-war road bridge in the UK (1955).

One steel bridge in Wales is famous for a very different reason: the Cleddau Bridge (SM 975045). It collapsed during construction in 1970 and the subsequent inquiry produced a new design code for steel box-girder bridges. Box-girder bridges had been built in the 19th century in wrought iron and the concept was revived in steel in the 1960s for large span bridges using high-strength steel and welded construction with an aerodynamic cross-section. In 1968 a contract was let for a new steel box-girder bridge across Milford Haven from

37 *Newport Transporter Bridge.*

38 *Queensferry Bridge. (© NH)*

39 and 40 *The Cleddau bridge. The original collapse was over pier six, the first pier from the right. (© Flint & Neill and ICE).*

Pembroke Dock to Neyland to replace a ferry. Developments in the Haven had created a need for better communications between both sides. The bridge was due for completion in 1971. However, in June 1970 a section of the bridge collapsed during the launching of a prefabricated section of the deck. In December a box-girder bridge over the Yarra River in Australia collapsed during construction and one in 1971 at Koblenz. The causes of these collapses were different in each location but an urgent review of the design methods was necessary. The Merrison Committee of Inquiry into the Milford collapse concluded that the current design codes were inadequate for this new form of bridge and a new set of design rules were produced to cover the design. After redesign the work recommenced and the bridge was finally opened in 1975. This Welsh failure therefore led to a worldwide review of design and new criteria for such large span structures.

Concrete Bridges

Concrete, like cast iron, is comparatively weak in tension, so the first concrete bridges were arches, using the material in compression. To overcome the poor tension characteristics of concrete, steel reinforcing bars can be added to the areas of the concrete structure which are in tension thus enabling the composite material to be used for beams and columns. Reinforced concrete bridges were developed on the continent by François Hennebique in the 1880s, and his techniques were brought to Britain under license by L.G. Mouchel. Mouchel had come to Wales from France in 1876, to Briton Ferry. He initially traded as a shipbroker and coal merchant. He was also the French vice-consul in South Wales from 1879 until his death in 1908. Hennebique persuaded him to act as agent for his new system and Mouchel's engineering practise developed from this. Between 1898 and 1907 he was involved in over 20 projects in South Wales for reinforced concrete bridges, coal handling hoists, jetties, stores and other buildings and many more in England.

As well as the use of this new technique for bridges it was adopted rapidly for other forms of construction. Weavers Mill in Swansea, now demolished, was the forerunner for the use of 'ferro-concrete' for multi-storey framed buildings when constructed in 1898 with the Hennebique patent by Mouchel in conjunction with H.C. Portsmouth, a Swansea architect. The flour mill, rectangular in plan, measured 80ft by 40ft and was 112ft high. The building was the first to use 'bent-up' reinforcing bars to resist diagonal tensile stresses. The building was demolished in the 1980s to make way for development of the disused dock frontage, it being the last remaining dockside building in the area and not particularly photogenic. A section of the reinforced concrete from the building has been retained on site.

Berw Road Bridge (S22) in Pontypridd is an early Mouchel bridge and was the longest of its type in the UK when built around 1905. By 1910 bridges of the new material were becoming commonplace. Reinforced concrete became almost the material of choice for new small- and medium-span bridges.

The development of prestressed concrete in the mid-20th century enabled longer spans to be built in concrete. Prestressed concrete is produced by casting the concrete around steel bar or wire reinforcement that has been stretched and held in tension while the concrete sets. The tension is then released and the concrete is compressed by the contraction of the reinforcement. This enables higher loads to be applied before the concrete reaches the tension stage when the steel reinforcement takes over the tension load. Longer spans are possible and less materials required, reducing loads on supporting piers and foundations which themselves can be smaller.

As with reinforced concrete, the first prestressed concrete bridges in Britain were based on continental practice. The initial impetus came during the Second World War

41 *Weaver's Mill, Swansea. (© Owen Gibbs Collection)*

when it was expected that significant numbers of bridges might be destroyed by enemy action, and a stock of pre-cast, prestressed beams was made for emergency use. In fact, only two such bridges were needed.

A further development was post-tensioned concrete, where the steel bars or cables are contained in ducts cast in the concrete and tensioned after the concrete has been cast by jacking the bars or cables against anchorages cast at each end of the concrete unit. Once the tension has been applied to compress the concrete the ducts are grouted to protect the stressing cables. In 1948 the first *in-situ* post-tensioned bridge was completed at Fishtoft near Boston with a 72ft clear span. The longest span in Britain in prestressed concrete is 623ft at Orwell Bridge, near Ipswich, Suffolk.

The comparative lightness of prestressed and post-tensioned concrete compared with reinforced concrete enabled designers to provide very elegant structures. Combined with modern design tools it has also made possible new techniques of construction. Balanced cantilever methods (whereby the bridge is built in increments each side of a support pier) were used to good effect in 1964 at Medway Bridge, which has a main span of 500ft. Construction can be either by casting the increments *in situ* using specially designed shuttering to form the hollow concrete sections or by casting the sections off-site and transporting and lifting into place. The method is particularly useful where access to the land under the bridge is difficult. The **Dee Viaduct (N44)** (A483 Newbridge Bypass) was constructed using the *in-situ* method. Off-site casting was used for the glued segmental viaducts in Cardiff for the Southern Distributor road and also for the approach spans at the Second Severn Crossing.

42 *Dee Viaduct. (© Welsh Assembly Government)*

Examples in Wales include the three-span Taf Fawr (1964) (SO 028079), with a 216ft centre span (here in 1981, some 17 years after construction, severe corrosion was found in the prestressing cables and the deck had to be completely replaced, again by the balanced cantilever method) and the five-span

Dee Viaduct (1990), with 272ft central span, mentioned above, the highest trunk road bridge in the UK, shown here under construction.

A bridge just south of the Dee Viaduct on the A483 Chirk Bypass (SJ 299374) was constructed by an alternative method, 'cast push' (or incrementally launched). The bridge was cast in short lengths on the southern bank and the section pushed out over the valley using hydraulic jacks. As one section was pushed out another was cast behind and stressed to the previous one before it itself was jacked outwards. Permanent and temporary support piers had been constructed in advance in the valley. The process continued until the bridge deck was completed across the valley. The bridge was then finally stressed to its design load. This project was particularly difficult as the bridge is not straight and level but curved in both plan and elevation. Its construction is a particularly noteworthy example of the skill and expertise of the contractor concerned, Christiani & Neilson.

Suspension Bridges

For longer spans, cable-stayed and suspension bridges are the rule. In suspension bridges the deck is supported by hangars from a suspension cable, in turn supported by towers. The deck of a cable-stayed bridge is directly supported by a fan of cables from the towers to the deck.

Early suspension bridges in Wales are the well-known Menai and Conwy bridges built by Telford in the 1820s (**N4, N8**), Menai being the longest in the world when built, with a total length of 1,050ft and a clear span of 580ft. Telford led the world with his aqueduct at Pontcysyllte and continued to extend the boundaries of civil engineering with this huge suspension span.

Captain Sir Samuel Brown was an early proponent of chain suspension bridges. He originally developed iron chains for the Royal Navy to replace the hempen cables then in use and established a chain works at Millwall in 1812. He considered establishing works in every port but eventually settled on building his second at Newbridge in South Wales (later known as Pontypridd). This Brown Lenox works produced the chains for Brunel's '*Great Eastern*'. However, he also saw the possibility of using chains for bridges. He had collaborated with Telford in a proposal to cross the Mersey in 1816, which did not proceed, and patented his design of suspension chains made up from flat eye-bar links in 1817. Telford used this system for his bridges at Conwy and Menai. Telford had initially considered one-inch-square wrought-iron rods welded together as a continuous cable, but difficulties in ensuring the quality of the welding led him to select the flat chain option.

Brown's first large suspension bridge was the Union Bridge over the Tweed built in 1820, a span of 361ft which still stands today. He also built the chain pier at Brighton with four 255ft spans suspended from cast-iron towers in 1823. He was generally recognised as the most experienced builder of suspension bridges in the country. He built two in Wales, at Llandovery and at Kemeys Commander, Usk (SO 356056). The latter built in 1830 was replaced in 1906 with a different design but is still referred to as the Chain Bridge. The bridge at Llandovery built in 1832 was replaced in 1883.

One of the other innovative chain-bridge designs was that patented by James Dredge of Bath, a bridge design that looks like a 'web of iron'. A brewer by trade, Dredge turned his hand to developing a new form of suspension bridge with the help of his blacksmith brother. The result was a hybrid development, a chain-bridge with a catenary chain that tapers from the top of the towers, the number of links reducing towards the centre. Each inclined suspension rod was joined to the links in the suspension chain, which helped to resist twisting and could be built up of many small components.

43 *Union Bridge, Horncliffe. (© Owen Gibbs Collection)*

His first bridge, the Victoria Bridge, was built in Bath in 1837 and remains in use as a footbridge to this day. He was responsible for approximately fifty bridges of this type, in Britain and abroad. A number of his bridges are still standing with others built under licence or copied from his designs. Several were built in Wales but have been demolished, except for one at Doldowlod (**M13**) on a private estate.

There is another old suspension bridge in poor condition near Llangollen, at Llantysilio, crossing the Dee (SJ 199432). The original bridge was built about 1818 but has been rebuilt twice, the last in 1929. It is believed that the 12 chains still in place are the original ones. It was built by a local business man, Exuperius Pickering, to bring coal across the river to his wharf on the canal. Originally built as an under-slung girder, it

44 *Victoria Bridge, Bath. (© SKJ collection)*

was rebuilt as a conventional suspension bridge after flood damage.

Major 20th-century bridges in Wales include the Wye Bridge (1966) (ST 540912), a 1,340ft-span cable-stayed steel bridge which crosses the Wye below Chepstow and leads to the Severn Bridge (1966) just over the border. At 3,240ft it is one of the longest suspension bridges in the world and now a Grade-I listed structure. The bridge sections were fabricated nearby in Chepstow, floated out and lifted into position under the suspension cables. The centre section of the Second Severn Crossing (1996) is a cable-stayed 1,490ft-span steel bridge built by a balanced method whereby sections were floated out on barges and lifted into place to form balanced cantilevers each side of the support towers. **George Street Bridge (S15)** in Newport, with a 500ft span, was the first concrete cable-stayed bridge in the UK (1964).

45 *Llantysilio Suspension Bridge.*

VIADUCTS AND AQUEDUCTS

Most railway viaducts and canal aqueducts are built of masonry with the notable exception in Wales of Pontcysyllte, which has a cast-iron trough supported on masonry piers, and the now demolished **Crumlin Viaduct (S16)**, built in 1857 using wrought-iron trusses and piers. Walnut Tree Viaduct at Taff's Well near Cardiff was formed of seven steel lattice girders. It was built in 1901 by the Barry Dock & Railway Company to cross the valley. The B D & R was one of the last private lines to be built and was forced to use considerable engineering work as the more favourable routes had been taken by earlier railways. At 120ft high and 1,548ft long this spectacular viaduct was designed by Sir James Szlumper. It was demolished in 1969 though two of the massive brick piers remain (ST 128825). **Chirk Aqueduct (N40)** (1801), is a hybrid of masonry and iron in that Telford used cast-iron plates to form the channel, reducing the amount of masonry required to support it.

The early viaducts of the railway era were often of timber but these were soon replaced with more durable masonry, although one at Aberdare survived until 1947. **Barmouth Viaduct (N31)** with 113 spans carrying the Cambrian Coast Railway over the Mawddach Estuary is a fine existing example of a timber construction. It was built in 1867 although the original lifting drawbridge span was replaced in 1901 with a two-span swingbridge in steel, which has not moved for many years. The Loughor (Lwchwr) Viaduct (SS 561890), built by Brunel on the South Wales Railway, still has timber trestles supporting a later steel deck. Brunel's original timber trestles have been strengthened and augmented but the lower sections seen at low tide are the original timbers.

The multitude of railways crossing the South Wales valleys led to the construction of many high-level viaducts. Most of these have been demolished as lines closed but some have been retained long after the railway had gone and serve a new purpose as footways and cycle paths. All are significant features in the landscape.

Important examples include the 1866 **Cefn Coed Viaduct (S26)** at Merthyr Tydfil, and the 16-arch curved **Hengoed Viaduct (S18),** now part of National Cycle Route 47. Brunel's masonry viaducts at Newbridge (Pontypridd) (ST 071900) and **Goitre Coed, Quakers Yard (S20)** on the Taff Vale Railway built in 1841 are still in use but his 1850 timber viaduct at Landore, near Swansea, on his South Wales Railway was replaced in masonry and wrought iron in 1889 (SS 663959).

The masonry viaduct at Goitre Coed has unusual octagonal piers. The viaduct crosses the river on a large skew angle and Brunel was concerned about the possibility of scour. He decided on octagonal piers to ensure they were square to the river flow offering the least resistance. The viaduct is therefore a skew viaduct but without conventional skew arches.

There are many other viaducts still in use on railways throughout Wales. Some of the notable ones include those built by Stephenson as part of his **Chester & Holyhead Railway (N9)**.The **Ogwen Viaduct (N11)** has 22 spans and was built on this line in 1848, as was the 13-arch viaduct at Penmaenmawr and the 19-arch **Malltraeth Viaduct (N14)**. Railway viaducts at Chirk (**N42**) and Cefn (**N43**) both built in 1848 on the Shrewsbury & Chester Railway as well as the 1868 **Cynghordy Viaduct (M14)** on the Central Wales Line are still in current use. **Knucklas Viaduct (M11)** on the Central Wales line has 13 50ft spans and was built by the Central Wales Railway in 1863. The 1879 Lledr Viaduct, **Pont Gethin (N25)** (SH 780539), carrying the Conwy Valley line also remains in full use.

Significant aqueducts still in use in addition to Pontcysyllte include Chirk (**N40**), in North Wales, Brynich (**S12**), south of Brecon, and several on the Montgomeryshire

46, 47 *A model of Landore viaduct produced at the time of demolition is in storage at Swansea Museum.*

48 *Cowbridge viaduct. (© Owen Gibbs Collection)*

Canal. There is a small cast-iron aqueduct at Resolven built in 1851 to carry the Rheola Brook over the Vale of Neath Railway (SN 831027). Brunel, the engineer for the line, may well have been involved in its design although this cannot be proved. The cast iron was from George Kennedy's foundry in Bridgwater.

One aqueduct which became a railway viaduct and later a road viaduct is Pont Rhyd y Fen (SS 796942). Built in 1827 to carry a water course over the river at a sufficient head to power a water wheel at Oakwood ironworks near Cwmafan, it had a railway laid across it in 1841 and when the railway closed in the 1950s was converted to a highway. It is now restricted to light traffic only.

A concrete viaduct built in 1965 carrying the A48 bypass to Cowbridge in South Wales has 15 spans and was an early use of prestressed concrete. Prestressed beams were cast on site and then erected onto *in-situ* reinforced concrete columns and a reinforced deck cast over the beams to complete the structure (SS 995750).

TUNNELS AND EARTHWORKS

The railways and canals of Wales have required many major engineering works to cope with the considerable obstacles of topography to the construction of suitably graded layouts. Pontcysyllte Aqueduct has already been mentioned, but the canal embankment immediately south-east of it is also a monumental engineering achievement, somewhat overshadowed by its famous neighbour.

Wales has the longest single-track land railway tunnel in Britain at **Blaenau Festiniog (N25)** carrying the Conwy Valley line which runs up the Lledr Valley in Snowdonia and through to Blaenau Ffestiniog on the south-west of the Snowdonia Range. The tunnel is 2 miles 340 yards long and was built in 1879 by the LNWR. The rock through which it was being driven was extremely hard and the first contractor failed to cope. It was not without its challenges for the engineers. Because of its length it was being worked from three shafts and from each end in eight separate headings. All met almost exactly to form the final bore.

There are many other tunnels and earthworks throughout the Principality where the topography challenged the ingenuity of the civil engineer. One on the Merthyr, Tredegar & Abergavenny Railway built from both sides of two valleys on a very tight curve and high level met precisely. The **Severn Tunnel (S10)** was the longest underwater tunnel until the Channel Tunnel was built and is of course partly in Wales. Britain's first immersed tube tunnel was opened in 1991 carrying the new A55 dual carriageway under the Estuary, bypassing the World Heritage Town of Conwy **(N18)**.

The rocky headlands between Conwy and Bangor presented obstacles to travel for hundreds of years until they were tunnelled in the 1840s for the Chester & Holyhead Railway and in the 1990s for the A55 expressway, with tunnels at Penmaenbach and Pen-y-Clip. Telford had chosen a route across the headlands mainly on ledges about

49 *Pont Rhyd y Fen, 1980. (© Owen Gibbs Collection)*

halfway up. This route was abandoned when the road was improved in the 1930s – a wider road constructed around the headland, again on rock ledges with short sections of tunnel – but it has been brought back into use as part of the A55 development in the 1990s to provide an alternative route for walkers and cyclists who are not permitted to use the two new tunnels through the headlands.

In Mid Wales there are also several interesting engineering features dictated by the mountainous terrain. At Friog the coastal section of the Cambrian Railway near Fairbourne is cut into an artificial shelf 100ft up the sea cliff and an avalanche shelter has been constructed. After two serious derailments due to landslides, the latest in 1933, a concrete shelter was constructed by the GWR in 1936 (SH 607116) to protect the line. Wales also has the deepest rock cutting in Britain at **Talerddig (M10)** between Newtown and Machynlleth on the Cambrian Railways, 120ft deep and 400 yards long, constructed in 1863.

PORTS AND HARBOURS

In Wales, with the exception of the Wye on the English border, the rivers generally were not suitable for inland navigation. The Severn Estuary was used for coastal traffic with numerous small river wharves near river mouths from Chepstow to the far west. There was no navigation generally above the tidal reaches. For example the River Neath was navigable only to Aberdulais, the Tawe to Morriston, just above Swansea, the Loughor to Pontardulais and the Towy to Carmarthen. There are also few natural harbours other than Milford Haven. The pattern was similar around the coast from Cardigan northwards and along the North Wales coast to the Dee Estuary. In Roman times the Dee was improved to provide access to Chester and in medieval times the River Clwyd to Rhuddlan (**N33.1**) was also improved but little was done elsewhere. Around the coast most tidal inlets had some form of quay for local and longer-distance trade. As well as Chester the Roman influence is evident at several places in the Severn Estuary, notably the Legion base at Caerleon just upriver from Newport.

Maritime evidence for coastal and continental trade has come from discoveries in the Gwent Area. The earliest vessel, or at least parts of it, found has been dated to the Bronze Age, *c.*1000 B.C. Remains of a Romano-Celtic ship were found near Magor and in 2002 a medieval vessel was found during excavations for a new Riverside Theatre in Newport, with a probable construction date of the mid-15th century. Remains found with it suggest trade with the Iberian Peninsula.

Over 60 ports and havens have experienced greater or lesser importance in the history of Wales. Iron-Age settlements on the South Wales coast traded with Brittany and the Mediterranean. Roman garrisons from Caerleon to Holyhead were supplied and reinforced by sea, and Edward I in medieval times built his castles in the north and west where they could be supplied and reinforced by sea.

Documents from the 13th century record ships at Kerdiff (Cardiff), Newport and Bristol. In the late 13th century the port dues were agreed as part of the Cardiff Borough revenues. In 1610 Cardiff was described as 'a quiet fishing port' and it remained so into the early 19th century.

Reforms of Henry VIII encouraged the search for minerals and the formation of chartered companies for manufacturing. Many Welsh ports expanded. Iron and non-ferrous metal smelting was boosted. Neath followed Swansea in the 1580s with copper smelting, using local coal and ores from Cornwall and Anglesey. Wire-making works developed on the Wye at Tintern. About 1560 customs control was established in Wales with Milford as head port and secondary ones at Cardiff (controlling the coast from Chepstow to Milford) and Chester (controlling Barmouth to the Dee). Port books are extant from the late 16th century for Cardiff, Milford and Chester which enable the coastal and international trade of the period to be assessed.

Coal mining led to maritime trade whenever shallow seams were found near the coast or outcropping in valley sides, often associated with limestone and ironstone deposits. This applied westward of Neath and particularly the Carmarthenshire coast west of Swansea. Coal was known to be worked at Margam as early as 1250. Transport was by packhorse so ports on the coast near mines thrived. There was considerable maritime trade and in 1599 Swansea recorded 47 shipments, all but two carrying coal; Neath had a similar amount that year. Small amounts of barley and dairy products were also shipped out. Cardiff records in 1579 include eight foreign vessels bringing salt, iron and wine from France. In the 17th century coal was being shipped from Pembrokeshire to London, mined from outcrops. Its use was mainly to fuel lime kilns.

Trade increased tenfold in the 17th century with total cargoes in and out of Wales of some 200,000 tons a year. In 1700 Swansea alone had 100 ships in total bringing copper in and taking coal out. Outside the coal-mining areas Chepstow and the Wye were exporting timber from the Forest of Dean for shipbuilding and bark for tanning to Bristol; Newport was exporting bark to Ireland; Monmouth was importing salt and wine from France and exporting skins to Bristol. Ports in the Vale of Glamorgan were sending dairy products to England. Carmarthen was importing barley corn and malt and exporting large amounts of Welsh flannel. Aberystwyth was importing coal for smelting lead and silver and also exporting lead ore. Pwllheli and Flint were exporting lead and lead ore and copper ore was being sent from Anglesey to Swansea.

In North Wales coastal trade was mainly to Lancashire. In the 16th century there was already a small trade in slate using small 20-ton vessels. The export of slate grew rapidly as packhorse and pannier were replaced by tramways. From 1800 the number of shipments increased dramatically, with exports going to Europe, America and the West Indies as well as to England. By 1850 over 200,000 tons of slate was being exported each year, ships returning with foodstuffs, olive oil, grain and phosphates. Other industries in North-East Wales used port facilities along the coast and also canal and road transport to England. Lead has been mined and smelted from ancient times in Flintshire. Coal has been shipped from Mostyn from the 15th century and a formal quay was in being as early as the 17th century before the docks were constructed in the 1840s. Early foundries are known in the Rhuddlan area.

With road improvements in the mid-18th century and the development of iron-works on the northern edge of the South Wales Coalfield the coastal ports from Newport to Swansea began to grow, considerably boosted by the new canals in the late 18th and early 19th centuries. There was a corresponding decline in the ports west of Swansea.

The more favourably located river wharves expanded and the use of others declined. But these facilities were soon inadequate. The commercial pressures encouraged the formation of Dock and Harbour Authorities to provide improved and non-tidal facilities. Dock development sometimes began with the creation of a half-tide basin by constructing a new cut in the river and isolating a suitable bend for conversion to a wet dock, that bend usually being already in use for river wharfs. Lock gates were added to maintain water levels. As trade grew further and vessels increased in size, purpose-built docks were added further downriver. Dock construction continued to migrate further seawards to deeper water and in most ports the later docks were built on reclaimed tidal salting and mudflats. Ports continued to expand until the early 20th century.

The final development as ships became too large for the dock locks and the water depth in the docks was the offshore jetty for loading and discharge of bulk cargoes. The transport of oil imports in larger and larger ships produced the need for deep-water berths. Milford Haven in particular grew rapidly in the 1950s with several deep-water oil terminals as the oil terminal in Swansea Docks could not take super-tankers. At Port Talbot a deep-water terminal for the import of iron ore was built in the 1960s, replacing in-dock facilities at Newport and Port Talbot.

The earliest docks were at Neath, Pembrey, Burry Port and Llanelli. Dock and railway companies were formed in Newport, Cardiff, Barry, Porthcawl, Swansea and Llanelli. Harbours and quays were constructed or extended along Cardigan Bay, including Cardigan, Aberaeron, Aberystwyth, Aberdovey and Porthmadog, on Caernarfon Bay at Felinheli (Port Dinorwig) and Caernarfon and on the North Wales coast at several places including Holyhead, Penrhyn and Mostyn. Amlwch on Anglesey was developed to export copper ore from Parys Mountain.

Some growth occurred to service the coal and lead mines and ironworks of Flintshire and North-East Wales, including Foryd Harbour near Rhyl and Talacre Harbour. Small jetties were built or enlarged at Connah's Quay and Queensferry. But major ports did not develop to serve the coal and iron industries there because of the proximity of the Mersey ports and the ease of west-east communications.

The major South Wales ports developed around the export of coal and tinplate and the import of iron ore as local sources were used up. Swansea in the early 18th century was importing non-ferrous ores for its extensive smelting and refining industries. The growing worldwide demands for steam coal for locomotives, ships and industry made South Wales one of the key areas for expansion. Its steam coal was probably the best in the world at the time, having a high calorific value. The demands for iron products, and later steel in sheet and bar form, fuelled the rapid expansion of the coal and iron industries and the associated infrastructure. Much of the ironwork for the new railways in Britain and abroad was produced in South Wales ironworks.

Many of the steelworks migrated to the coastal areas in the late 19th century to avoid the need to haul ore uphill and for the easier downhill haulage of coal. The one exception was at Ebbw Vale, which remained in full use until the late 20th century with ore being imported through Newport Docks and taken by rail to the head of the valley. As the coal trade declined after the 1940s the ports were redeveloped for general cargo, bulk timber and containerised cargoes, and the old coal-handling facilities were converted in the 1950s and 1960s to general cargo quays. These have been concentrated on the parts of the dockside suitable for the largest vessels, and in the last few decades as trading patterns have changed large parts of the older docks estates have been redeveloped for housing, leisure use and industry. Some of the older docks have been filled and developed and little remains to be seen of the original infrastructure apart from the water areas and quaysides.

50 *Entrance to Penarth Marina.*

Docks and harbours have also been developed as marinas and moorings for leisure craft and still retain the original quayside features, although much of the associated infrastructure has been lost. Penarth Docks and the South Dock in Swansea have become marinas with associated facilities and quayside housing, as have other old ports around Wales.

At Conwy a major marina was developed, not in a disused harbour but in the huge man-made excavation for the construction of the six 30,000-ton concrete units for the **Conwy Tunnel (N18)** in the 1990s. The six units were cast in a dry excavation which was then flooded and the units floated out in turn to be sunk in a prepared trench in the river, forming the tunnel. The resulting redundant excavation has been converted into a leisure marina.

DRY DOCKS AND SHIP REPAIR

Maritime trade is obviously dependent upon ships and these have to be built and repaired. As well as the construction of docks, civil engineers were needed to design and build repair facilities, slipways and graving, or dry, docks.

In Wales from early times various river estuaries developed facilities for the construction of timber vessels. These yards were widespread around the Welsh coast but were mainly for small coastal shipping. Some locations developed as the quality of their work became widely known and small and large industries grew where this expertise and sources of suitable timber were found together. Ships were built and repaired on slipways and purpose-built grid-irons or 'hards' alongside or part of river wharfs and also within the wet docks themselves. Ships could be berthed on a grid-iron for repair or manoeuvred into a cradle at high water which would then be pulled up a slipway to above high water where they could be repaired. Yards were often associated with sail lofts, block workshops and rope making.

51 *Channel Dry Docks, Cardiff, 1899. (© Stewart Williams)*

As ship sizes increased, the demand for larger repair facilities grew and this was stimulated further with the development of iron ships and steam propulsion and the need to handle heavier loads. The new wet docks being built in the early 19th century often included the construction of purpose-built dry docks, with facilities also developing separately near the dock entrances in the estuaries or lower sections of the rivers. Such facilities outside the docks had the advantage of not incurring dock dues.

Entrances to the dry docks were usually sealed by conventional lock gates or iron caissons which could be floated and moved aside for vessels to be berthed in the dock. One early example is Telford's Graving Dock in Holyhead (**N6**), built in 1825, originally tidal but within a few years pumping facilities were added to drain the dock and increase its accessibility and use. Significant ship repair facilities developed at the large South Wales Docks. There are many examples around the country and in many places the old slipways and dry docks can still be found.

Ships up to 500 tons were built in the 19th century and probably earlier on the Wye at locations from Chepstow to Monmouth. The yard at Chepstow developed into a major facility building ships up to 6,500 tons into the 20th century. At Cardiff on the River Taff Estuary there was the Bute Shipyard, for a time managed by the Scott Russell family, who built at least one iron ship there. The attempt to establish Cardiff as a centre for shipbuilding using Dowlais iron did not succeed and Cardiff became instead a major ship repairing centre. Neath Abbey Ironworks was to pioneer the building of iron-hulled screw-propelled ships as early as 1846 at the Neath Abbey shipyard. Shipbuilding and repairing developed on the Tawe and continues today.

Pembroke Dock became a Naval Dockyard in 1814, taking over from a small earlier one on the other side of the Haven which was founded in 1796. It became one of the Navy's important shipbuilding centres with over 260 ships being built over its 112-year life. Between 1930 and 1959 the dockyard was used as a base for the RAF Sunderland flying boats. Several of the original docks survive as well as two 1930s aircraft hangars, the last of their type in the UK. Smaller private shipyards were also established on each side of the Haven.

52 *Windsor slipway, Cardiff, 1899. (© Stewart Williams)*

Around the west coast small ships were built at most harbours. Fishing smacks and schooners up to 180 tons were built at Aberaeron, New Quay, and Aberystwyth. Where suitable harbours did not exist ships were built on the river banks or the beach. For example at Llanon and Llansanffriad on Cardigan Bay ships up to 50 tons were built on the shore. Cardigan, which had become one of the more important ports on this coast by Elizabethan times, had a shipbuilding industry during the 17th century and was building ships from 30 tons up to 160 tons from the late 18th century until the end of the 19th century, although for the latter part it was mainly repair work as sailing ships were gradually being replaced by steam-powered iron vessels. Over 140 ships were built there.

Barmouth developed around shipbuilding. Porthmadog became well known for the quality of its ships. Shipbuilding at all these locations peaked around the mid-19th century as iron and later steel steam-powered ships became predominant. Some yards continue boat-building for the leisure industry to the present day. For example one of the three 19th-century shipyards at Bangor developed to build ships for the carriage of slate still exists as a yacht builder. Ships were built at harbours on Anglesey for the shipment of Parys Mountain copper ore from Amlwch.

53 *No. 1 Graving Dock, Mountstuart Dry Docks, Cardiff, 2009.*

54 *No. 3 Graving Dock, at Mountstuart Dry Docks, is now used for small fishing vessels.*

In North Wales favourable locations for shipbuilding were concentrated to the east and the Dee Estuary with yards at Flint and later Connah's Quay, building wooden ships into the early 20th century. The iron steam clipper *Royal Charter* was launched at Sandycroft on the Dee in 1857, a ship that would be wrecked two years later off Moelfre, Anglesey, with great loss of life, in a storm that has gone down in history as the Royal Charter storm. A yard at Queensferry opened in 1885 built iron vessels and barges up until 1925 when the trade moved to the larger shipyards outside Wales with better facilities for the increasing size of ships and the new diesel-powered vessels.

DOCKS IN SOUTH WALES

The major docks in South Wales were and still are Newport, Cardiff, Barry, Port Talbot and Swansea. Smaller docks existed at Porthcawl, Briton Ferry, Llanelli, Burry Port and Pembrey. These together with older parts of the major docks have closed over the years as vessel sizes increased or trade declined. Much of the civil engineering for these docks is still visible, in the working areas of the current docks and in the redeveloped areas of the older parts.

Newport in the late 18th century, despite having a population of less than 1,000, was a busy port. In 1791 for example it had over 250 ships with a total registered tonnage of 12,000 tons clearing the port, using river wharves on the Usk, and this before the completion of the canal. As trade grew the need for improved facilities was necessary. The first wet dock in Newport was built in 1842, by which time the town's population had grown to over 15,000. This dock had the largest dock gate of its time at 64ft wide. This was intended to attract Brunel's *Great Western*, which was using the estuary of the River Avon. Newport hoped to attract the trade and the *Great Western* to the town but the GWR chose Liverpool instead as at the time there was no direct rail link to Newport. This dock was followed by the North Dock (28 acres) built further downriver in 1875 for coal exports and timber imports with a lock to the River Usk.

The South Dock was added in 1893,, also with a lock to the Usk, and the old east lock was used as a dry dock. Capacity needed to be increased further and in 1907 and 1914 the South Dock Extension (96 acres in total) was constructed with a new 1,000ft-long, 100ft-wide, lock to the Usk near its mouth. This massive lock, the largest in the world at the time, was so large that it was not until the 1950s that the size of merchant ships exceeded its capacity. The extension was built partly by constructing a new cut to divert the River Ebbw to the west, using the bypassed length as part of the new dock extension. Transit sheds were built on the south side for general cargo imports and steel and tinplate exports while the north side was used for coal exports. Iron ore was also imported via the South Dock for Ebbw Vale steelworks and others in the area and post-1950 for the new Llanwern steelworks east of Newport until 1970, when iron ore imports moved to the new tidal harbour at Port Talbot. A new deep-water terminal for very large bulk carriers had been considered in the Severn Estuary at Newport but it was eventually concluded that a single facility at Port Talbot with ore transported by rail to the Llanwern Steelworks was a more viable option.

55 *Coal facilities, South Dock, 1926. (© Associated British Ports)*

In the 1950s, as road transport began to displace rail, many of the quaysides which were served by rail only were flush-paved for road access to the quayside cranage at Newport and the other major ports along the coast. With the advent of container ships in the 1960s new facilities were needed and derelict coal-handling areas on the north side of the South Dock were levelled and a new quay built for this new trade as well as for the import of packaged timber, in container-sized bundles, from as far away as British Columbia.

River wharves on the Usk continued in use and one became a major ship-breaking yard in the mid-20th century. Another continued in use handling steel products. Several graving docks on the east of the Usk were used as sand and gravel wharves and for coastal container ships when they became too small for the repair of the increased size of ships.

The first purpose-built wet dock built in Cardiff was the Bute Dock, later the Bute West Dock, although the earlier Glamorganshire Canal which joined the River Taff south of the town had a sea lock which provided a berthing facility. Quays in the River Taff which ran alongside the west of the town became disused after the construction by Brunel of a river diversion lower downriver to create an area for his Cardiff Station on the South Wales Railway in 1840, which meant that shipping could no longer reach them. The Bute Dock was built to the east of the canal in 1839 and at the time was the largest masonry dock in the world. The first rail-borne shipments were made in 1844 but it was only a few years

56. *New bulk timber and container quay, Newport Docks, 1967.*

before the rapid expansion of coal exports necessitated the expansion of the docks. The Bute East Dock was built in stages between 1855 and 1859 with a locked entrance basin to handle bigger ships, the Bute Dock being renamed the Bute West Dock.

Additional capacity was needed again by 1874 and the Roath Basin (13 acres) was built further seaward enclosing a previous tidal harbour, followed by the Roath Dock (33 acres) in 1887. This dock included facilities for iron ore imports for the nearby steel-works. The final extension in 1907 was the Queen Alexandra Dock (52 acres) with a larger new sea lock and connection to the Roath Dock. This had a general cargo quay and transit sheds on the north side, cold store on the east side and coal handling on the south.

In 1913 Cardiff exported 10.5 million tons of coal, but this figure was never to be reached again. Exports fell to about five million tons by the mid-1930s. The peak of the coal export trade had passed. The Bute West Dock became disused and was closed in 1964 and filled. The East Dock continued to be used for general cargo, mainly coastal shipping as the water depth precluded its use by the larger ships. It closed eventually in 1970. The last coal was shipped from Cardiff in 1964 and coal exports transferred to Barry. Iron ore imports ceased on closure of the East Moors (Dowlais) steelworks in 1978. In the late 1960s old coal-handling areas on the south side of Queen Alexandra's Dock were converted for the import of bulk timber cargoes.

More capacity was provided by the construction of a new dock at Penarth on the west of the River Ely in 1865 (26 acres after the 1884 enlargement) and the development of Ely Harbour as a tidal coaling facility. It had been originally proposed by Brunel in the 1840s as part of the Taff Vale Railway. A separate company had to be formed to build and operate the dock to get round a legal agreement with the Bute Dock owners regarding the TVR's use of their docks. It was built by the Penarth Harbour Dock and Railway Company. This company was the successor to the Ely Tidal Harbour and Railway Company formed in 1855 with their tidal coaling staithes becoming operational in 1858. Both these developments were promoted by the TVR and other industrialists as an alternative to the Bute Docks in Cardiff. The dock closed in the late 1970s and was partially filled, but it reopened in 1987 as a marina and was extended and improved when the Cardiff Bay Barrage was completed in 1999.

Port Talbot Docks were developed in 1836 for coal exports when the first section was built with a lock to the River Afan. By the latter half of the 19th century the nearby dock at Briton Ferry by Brunel was not large enough for the shipping of the time and it was not economical to enlarge it. The dock at Port Talbot was extended in 1894-8. The dock had a new lease of life when the new steelworks at Port Talbot were built in the 1960s and the dock became a major importer of iron ore, over three million tons a year by 1968. However, the entrance lock dimensions limited ships to 10,000 tons and new iron ore mines were opening up across the world. It was only economic to bring ore from Canada, Australia and Africa in larger vessels so consideration was given to building a new larger lock for 40,000-ton vessels. This was eventually discarded in favour of a deep-water berth off-shore. A new breakwater a mile long was built, a new deep-water channel dredged and a new jetty built capable of taking 65,000-ton vessels, with capacity for expansion to 100,000 tons. The work started in 1966 and the facility was opened in 1970.

Swansea had several river wharves on the River Tawe as far upriver as Morriston serving the non-ferrous smelting works. The first impounded dock was formed in 1852 by constructing a new cut for the river and using the old course on the west for quays where earlier tidal wharves built in 1846 were converted to a wet dock by the addition of lock gates. This became the North Dock.

57 *A view of Swansea in 1852 from a lithograph by Newman & Co. The New Cut is in the centre and the North Dock to the right. (© National Library of Wales)*

58 *Port Talbot showing new breakwaters and deep-water jetty. (© Associated British Ports)*

The South Dock (13 acres) also on the west bank of the river was built in 1859 and in 1881 the tidal Port Tennant on the east side of the river was converted into the Prince of Wales Dock (28 acres after the 1889 extension). In 1890 the Kings Dock (70 acres) was built south of this dock with a new sea lock for larger ships. In 1920 the development of oil imports led to the construction of the Queens Dock (104 acres) south of the Kings Dock with storage facilities for oil imports, piped to the refinery to the north-east at Llandarcy, and the export of refined products. Oil tankers continued to increase in size and eventually the import trade moved to deep-water berths at Milford Haven with new refineries there. The facilities at Swansea were then used for coastal shipping and storage of products from these refineries. The South Dock closed in 1969 but re-opened in 1982 as a marina with associated residential and leisure facilities. The National Waterfront Museum opened on the north side a few years ago. Further commercial and residential developments have been constructed around the Prince of Wales Dock on the east side of the Tawe.

One of the last docks to be developed in South Wales was **Barry Docks (S28)** (107 acres). The docks at Cardiff and Penarth, despite expansion in the late 19th century, could not cope with the amount of export trade, notably coal. Several coal owners led by David Davies of Llandinam decided to construct a completely new facility, as coal trains from their mines in the Taff, Rhondda and Rhymney Valleys could be days travelling the short distance to the docks due to the sheer volume of traffic and preferential treatment given to coal from mines owned by directors of the dock company.

David Davies had started out as a boy assisting his father on their farm and sawyer business but in 1848 branched out into minor road repairs and construction in the neighbourhood. He rapidly built a reputation for good quality work and organisation and in 1850, after working for the county surveyor, Thomas Penson, on Llandinam Bridge, obtained a contract for a new cattle market at Oswestry. Here he met Thomas Savin, a railway contractor, and jointly the partnership was responsible for several railways in Mid and North Wales until they parted company about 15 years later. Savin wished to speculate on the development of hotels in seaside towns now accessible by rail, an enterprise that led to his bankruptcy in 1866. Davies disagreed and they went their separate ways, Davies becoming a director of the Cambrian Railways.

Davies decided to invest in the new mining opportunities in South Wales and purchased land in the Rhondda Valley in South Wales. The enterprise was very successful and he and his partners formed the Ocean Coal Company. It was because of the difficulties with the Taff Vale Railway and the Bute Docks in access and charges to export their coal that Davies and his associates decided to build a new dock at Barry.

The Barry Dock & Railway Company was formed in 1884 and a new dock complex was built between 1884 and 1898 in the tidal inlet between the town of Barry and the island just offshore. The docks were designed by Sir J. Wolfe Barry and Brunel's son Henry Marc, who had produced earlier proposals for the docks. A new railway line was built northwards to intercept existing lines, the Taff Vale lines at Treforest and at Hafod in the Rhondda Valley. Later in 1901 a link across to the Rhymney Valley was added. In 1913 Barry exceeded Cardiff's coal exports of that year, shipping out just over 11 million tons. The last coal was shipped in 1976 and the last coal hoist demolished in 1981. The derelict coal areas have been redeveloped as a commercial and residential area.

Most of these docks and harbours were associated with railway companies and, on the amalgamation of railways into the four larger companies in 1922, came into their ownership. The GWR became owners of ports from Newport to Fishguard. These came into public ownership on rail nationalisation in 1948 after the Second World War, the smaller ones at Llanelli and Burry Port were sold and in 1982 the remainder were privatised and passed into the ownership of Associated British Ports.

HOLYHEAD AND FISHGUARD HARBOURS

Two important harbours were built in the early 19th and early 20th centuries in Wales for ferry services to Ireland. The first was at Holyhead (**N6**) in the early 1800s when the small existing harbour was developed with new breakwaters, quaysides and docks. Most can still be seen today. Several famous engineers were involved. John Rennie designed the new Admiralty Pier built between 1810 and 1821, and extended the North Breakwater. Thomas Telford designed the South Pier and a new graving dock, built between 1823 and 1831, and James Rendell extended the North Breakwater to create a harbour of refuge between 1847 and 1856. Later developments were made by the LNWR in 1860, who also built a new inner harbour and larger graving dock between 1875 and 1880, their chief civil engineer at the time being William Baker. Between 1856 and 1873 the North Breakwater was extended by John Hawkshaw. The port area has continued to be developed, the most recent improvement being for the high-speed ferry service to Dublin.

In 1906 the GWR developed a new harbour at Goodwick near Fishguard for their Irish Ferry Route. Brunel had considered this location among other options on Cardigan Bay for his terminus, eventually choosing Neyland as a more economical option. The GWR was looking for a shorter sea crossing and considered the economics justified the new harbour. This was constructed by blasting away the cliff on the south to create an area for the port facilities and using the rock for a breakwater. The Neyland terminal was closed when Fishguard opened.

AMLWCH AND PORTHCAWL

The growth and decline of small harbours can be illustrated by examples in North and South Wales. Amlwch in Anglesey in the mid-18th century was described as 'just a cove between steep rocks' and was used as shelter for the Liverpool pilot boat and

59 *Amlwch Harbour in 1860. (© National Monuments Record)*

small local vessels carrying corn, butter and cheese. The rediscovery of copper at Parys Mountain in 1762 increased the use of the small harbour. Ore was brought by horse and cart and shipped in small vessels to Flint; 16 tons in 1770 grew to 2,200 tons a year later to Swansea, Neath and Flint. The harbour was extended in the 1790s with 20,000 tons of rock blasted away. By the mid-19th century 18,000 tons of ore were being exported yearly. But in the 1850s the railway arrived in Anglesey and the port declined rapidly.

The small harbour at Porthcawl in South Wales (originally called Newton) was improved with a new breakwater and connected by tramway to Maesteg in the Llynfi Valley in 1825 for the export of coal and iron. In 1864 the entrance was improved and gates added to the basin, and in 1866 a small wet dock was added, designed by R.P. Brereton, who had been Brunel's senior assistant. In 1845 the port had shipped 35,000 tons of coal and 21,000 tons of iron; by 1871 this had risen to 165,000 tons of coal. But new docks which had opened east and west at Barry and Port Talbot offered better and less exposed facilities. Porthcawl Docks declined rapidly and the dock closed in 1902; the inner harbour closed in 1906 and the outer harbour in 1911. The dock has been filled and a new promenade built as the town developed as a seaside resort instead.

DEVELOPMENTS IN 21ST CENTURY

As trade has declined at the smaller ports and harbours around the coast many have had a new lease of life as marinas for the growing leisure industry. In South Wales, Penarth Docks in Cardiff Bay, South Dock at Swansea and docks and harbours at Llanelli and Burry Port have all been developed as marinas for leisure sailing, as have harbours on Cardigan Bay. Further north Porthmadog, Felinheli (Port Dinorwig), Caernarfon and Conwy all have marina developments and there is scope for more.

60 *Amlwch Harbour in 1994. (© RCAHMW)*

61 *(Below) Porthcawl Basin looking seaward.*

62 *(Below) Porthcawl basin looking landward showing old dock entrance.*

63 *Milford Haven showing new LNG import jetty. (© Aurora Imaging)*

New dock facilities for the import of Liquefied Natural Gas are being built in Milford Haven and the larger ports continue to have a steady import and export trade.

PIERS

A seaside pleasure pier is an artificial structure, generally constructed at right angles to the shoreline, used for promenading, steamer embarkation and entertainments. A conventional pier is founded on cast-iron screwed piles, carries hollow cast-iron vertical or raking columns set in lines or groups braced horizontally and diagonally with wrought-iron sections or tie bars, supporting longitudinal wrought-iron lattice girders braced by transverse wrought-iron lattice girders. These are in turn braced diagonally with wrought-angle irons, surmounted by transverse timber joists carrying a deck of longitudinal timber boards set with narrow gaps for drainage, with side timber string courses carrying cast-iron railings, usually very decorative. The form of construction for seaside piers has also been used for commercial jetties. There is one of these of similar age at the former Royal Navy Mine depot at Newton Noyes near Milford Haven (now closed).

64 *The purpose-built casting basin at Conwy for the six tunnel units for the A55 tunnel, which was developed as a marina once the units had been placed to form the tunnel. This photo shows the basin after the first unit had been positioned in the river Conwy. (© Welsh Assembly Government)*

Piers were mostly a Victorian phenomenon, the first being the Georgian 1813-14 Ryde Pier on the Isle of Wight, the last being the Edwardian 1910 Victoria Pier at Fleetwood, Lancashire. Piers were associated with the health benefits gained from 'taking the air' at the seaside over the sea itself when the tide was in, together with the escapist concept of holidays. The escapist theme was accentuated by the foreign and exotic architectural styles (Moorish, Indian, Chinese, etc.) associated with the kiosks and theatres on the piers. Most piers had an enlarged pierhead for steamer embarkation. Initially most pier theatres were constructed at the pierhead to avoid local authority planning controls and taxes, but the closure of this legal loophole led to later pier buildings being constructed on enlarged entrances.

The foremost designer of seaside piers was Eugenius Birch, who designed 14 piers, including Aberystwyth Pier in Wales, five being built by the contractors Richard Laidlaw & Son of Glasgow. They built a total of seven piers, a total surpassed only by Head Wrightson of Thornaby-on-Tees, who built eight as main contractor, also completing two other piers. Over 90 pleasure piers were constructed, mainly around the coasts of England and Wales, with one or two in Ireland but none in Scotland. Just over half of them remain today as complete, truncated or derelict structures.

During the Victorian Era there was a boom in seaside pier construction in Wales and nine were built between 1845 and 1900, of which seven remain. These were generally of the typical standard format of cast-iron columns supporting wrought-iron beams and trusses with timber decking, the later Colwyn Bay pier using steel lattice girders. Piers were built at Beaumaris in 1846, Aberystwyth in 1865, Llandudno in 1877 (**N20**), Penarth in 1894, Bangor in 1896 (**N19**), Mumbles near Swansea in 1898 (**S36**), and Colwyn Bay in 1900 (**N21**).

The two that have gone were both on the north coast of Wales at Rhos-on-Sea and Rhyl. Rhos-on-Sea was unique in that it was the only seaside pier to have been built in one location then dismantled and re-erected at another location. It was originally designed and built by John Dixon in 1869 at Douglas, Isle of Man, where its length was recorded as being 1,000ft. The move to Rhos-on-Sea was also undertaken by John Dixon in 1896. The pier at Rhyl was designed by (Sir) James William Brunlees and built by Richard Laidlaw & Son in 1867. It was 2,355ft long but was damaged by a ship collision in 1883. The pavilion burnt down in 1901 resulting in the closure of the pier in 1913. Following repairs it was re-opened in 1930 but eventually closed in 1966, by which date the length had been reduced to only 330ft. It was finally demolished in 1972.

65 *Aberystwyth Pier.*

Beaumaris Pier started as a landing jetty of timber piles and iron girders in 1846. It was extended to 570ft in 1895 and a small pavilion added. Since the 1960s the town council and later the county council have carried out some restoration work. The pier still offers local cruises and fishing trips.

Aberystwyth Pier, 800ft long, was damaged by a storm in 1938 which resulted in it being reduced in length

66 *Penarth Pier.*

by a half. In 1977 this length was closed on safety grounds, only the pavilion being kept open. Restoration has been planned several times but not carried out.

Penarth Pier, 658ft long, suffered a fire in 1931 which destroyed the seaward pavilion and much of the timberwork. The deck was replaced but not the seaward pavilion and the full length is still in use. The pier was damaged by ship collision in 1947 and subsequently repaired by underpinning the cast-iron columns and constructing new reinforced concrete columns. There are plans to refurbish the pavilion at the land end but these are on hold pending availability of funds.

Mumbles Pier, 835ft long, was refurbished in the 1980s. A lifeboat station was added in 1956. Llandudno Pier, the most attractive and ornate of the Welsh seaside piers, extending 1,234ft from the headland, was extended back to the promenade later with a 45-degree bend to a total length of 2,295ft. A pavilion added in 1884 burnt down in 1994. This pier was unique in having a pavilion on land with the then largest swimming pool in Britain in its basement. A landing stage was added in 1891 and a pavilion at the seaward end in 1905. Bangor Pier, 1,500ft long, was refurbished in the 1980s and reopened to the public. Victoria Pier at Colwyn Bay, 316ft long, used steel in its construction. The original pavilion burnt down in 1923 but was repaired and the pier extended to 750ft. A new pavilion was added in 1934. The pier closed in the late 1950s but after a series of owners was reopened in 1995. However, the company refurbishing it has gone bankrupt and its future is uncertain.

LIGHTHOUSES

The use of warning lights for shipping dates back over 2,500 years to ancient times with beacons and land-based towers. Lighthouses are used for several purposes. The most obvious is of course to mark dangerous coastlines or hazardous shoals, but they are just as important in providing markers for safe navigation into harbours and ports. Initially they were unlit beacons, leading lights and day or sea marks often set on headlands as navigational aids. Early construction would have been simple timber and masonry structures. Later substantial masonry towers were constructed. In the late 19th and 20th centuries iron and steel were used to support the lanterns. The first lighthouses were land-based but wave-washed or rock-based lighthouses became necessary as the volume of shipping and ocean trading grew.

Early lighthouses used open fires as a light source, which obviously presented problems in inclement weather; candles were also used in medieval times with suitable protection, particularly as harbour lights. From the 18th century often coal would be the fuel. The maintenance of open fires was a very manually demanding task. Not only did the fuel have to be hauled to the site but also manhandled or craned to the top of the structure and then loaded into the fire brazier. Ash had to be removed. For coal fires several hundred tons of coal would be used in a year. Coal-fired lights were in use in Britain until the 1820s.

The use of oil developed in the 19th century; whale oil was particularly favoured but other oils were used as well. Paraffin came into use in the mid-19th century, replacing gas which had been used in some locations, manufactured on site from coal. However, it proved hazardous except where it could be obtained as town gas for land-based lighthouses and the transport of large quantities of coal presented logistical problems. Electricity from steam-driven generators began to be used in the mid- to late 19th century but still had the logistical fuel supply problems. Oil-fuelled generators followed to supply power where connection to a mains supply was not possible. Bottled gas, particularly acetylene, became the fuel of choice where mains electrical supplies were unavailable. Standby generators covered any electrical supply failure. Solar power has now been developed and has been introduced where appropriate from the mid-1990s. The light source would be concentrated and directed usually by mirrors. In the mid-19th century the use of glass lenses and prisms was introduced and developed into the complex arrangements we see today.

In medieval times in Britain there were many locations where a system of harbour lights was used. Often these were provided as a light from a church or chapel tower as a service to local seamen. St Anne's Head in Milford Haven is believed to be one of these early lights, possibly 17th-century or earlier. However, the main impetus for an organised arrangement of warning and navigational systems arose in Tudor times. Early lights could be very ineffectual, there was no regulation of brightness nor location and many hazards were unmarked. Trinity House, the organisation that now manages many of the navigational aids around the English and Welsh coasts, received a Royal Charter in 1514 and in 1566 was empowered to set up beacons and warning lights.

Whether land- or rock-based the very locations of lighthouses presented the civil engineers of the time with significant challenges. Sites were remote and accessible only with great difficulty. Transport of construction materials presented major problems and the manning and servicing with supplies after construction was not easy either.

The first Trinity House lighthouse was constructed in 1609 at Lowestoft as the first of a series of lights along the coast, paid for by a levy on vessels leaving the local ports. Their first lightship was commissioned in 1732. At this time it was not uncommon for private landowners to be permitted to set up a lighthouse and to receive dues from shipping using nearby ports and harbours for its cost and maintenance. The first Eddystone light in 1698 was an offshore private enterprise. Other lighthouses and navigational aids were built by Local Harbour Authorities. The mix of Trinity House and private lighthouses which developed over the next 200 years was not reliable and in 1836 Trinity House was empowered to compulsorily purchase private installations. In 1967 Trinity House started the process of electrification of their lighthouses and has continued to modernise and innovate to the present day. Their last manned lighthouse was automated in 1998.

One of the earliest lighthouses in Wales was Flatholm in the Bristol Channel, built in 1738 after requests from Bristol merchants to Trinity House a few years earlier. Trinity House authorised the building of a private lighthouse which was lit by an open coal fire until 1820 when it was taken over by them and an oil lantern added.

There are 16 Trinity House lighthouses around the Welsh coast. Most are land-based but three are rock-based and sea-washed. There are also many harbour entrance lights built by various harbour authorities to assist navigation. These are often on the seaward ends of breakwaters. There are several disused lighthouses that have been converted for other uses, mainly residential.

The most southerly Welsh lighthouse is the one on Flat Holm, the most westerly at the Smalls off Pembrokeshire and the most northerly the Skerries off Anglesey. This is also almost the oldest, built in 1717 as a private venture and taken over by Trinity House in 1814. The oldest is St Anne's Head, Milford Haven, built in 1714, also a private venture, but rebuilt in 1841 by Trinity House. There is evidence of a possible medieval light at this point.

The Smalls and Skerries lighthouses were built to assist navigation on the route to and from Liverpool. The shipping channel in the Irish Sea became a major trans-Atlantic route when the port of Liverpool became the main UK port for western trade in the 19th century, with regular steam ship sailings as well as cargo vessels. These lighthouses marked important navigational features on this sea route. Two lighthouses on the North Wales coast at Point Lynas and Great Orme were built by the Mersey Docks & Harbour Board to aid navigation to Liverpool.

67 *(Left) Mumbles, a typical coastal lighthouse.*

68 *(Right) Porthcawl – a typical breakwater lighthouse.*

The first British rock-based lighthouse was the Eddystone Lighthouse, over 20 miles offshore, built in 1698 on a dangerous reef which had claimed many ships. It presented significant challenges to its builder. The present lighthouse is the fourth on the site. It was followed in the first half of the 19th century with more rock-based towers. One of the experts was James Walker, consultant to Trinity House and second president of the Institution of Civil Engineers, who built 29 lighthouses, several in Wales, including West Usk, Trwyn-du, and South Bishop. He remodelled the original private lighthouse at the Skerries for Trinity House and rebuilt the Smalls Lighthouse. West Usk was land-based, Trwyn-du wave-washed. His predecessor Daniel Alexander designed the first **South Stack Lighthouse (N7)**.

Three lighthouses have been selected by the Panel for particular note. These are Barry Docks Lighthouse (**S28**), located at the outer end of the entrance breakwater to the port and constructed by the Port Authority *c.*1890. It is a circular iron tower. **Whiteford Point Lighthouse (S37)** on the Gower is the only existing cast-iron wave swept lighthouse in Britain. Built by Llanelly Harbour Commissioners in 1865 to replace a wooden one erected in 1854 it is a scheduled ancient monument. It is no longer operational and can be visited at low tide.

South Stack Lighthouse (N7) at Holyhead was built in 1809 although a light was envisaged here as long ago as 1665. It is a circular masonry lighthouse typical of most around the coast. Access to the lighthouse was by an iron suspension bridge over a 90ft-wide chasm. The original bridge was built in 1828 and the deck replaced with aluminium in 1964. Until the bridge was built access was via a long series of steps down and up the ravine. Loads were carried by pack horse, actually a pack donkey called Tommy. This bridge was replaced with a new one in 1997.

WATER SUPPLY

The basic components of a modern public water supply are a source of water, a means to treat it to potable standard, a facility for storage to maintain a constant supply and a network to distribute it.

The development of water supply systems can be traced back to Roman times. Roman engineers developed distribution systems using aqueducts and pipelines in many places, Rome being the prime example. Bath-houses and a fresh water supply to public fountains were a major feature of Roman civilisation. Chester is one location where Roman lead pipes have been found and in Dorchester remains of a 12-mile-long aqueduct (SY 614955–674905) able to supply up to 13 million gallons a day can still be seen in places. However, it was not until late medieval times that public water supplies began to develop, replacing manual abstraction from a nearby spring, stream, well or village pump with water transported some distance from rivers and streams to supply an entire town or village. Chester is known to have had a system in the 16th century which raised water from the Dee by a water-driven pump and distributed it by pipework. This private undertaking became a statutory one in 1826.

As the industrialisation of Wales took shape in the late 18th and early 19th century one of the major needs was a water supply for industrial processes. Many of the ironworks were water-powered and had a need for a secure water supply. The natural sources of water from springs, wells and pumps and abstraction from lakes, streams and rivers which had been used for domestic purposes and rural industries such as woollen mills and papermaking were completely inadequate for the much greater demands of the new industries. There is little underground resource in Wales, as the geology provides few aquifers. It was necessary therefore to provide large volumes of water from other sources.

Initially some of the canal systems were available for conveying water but this also needed to be supplemented, particularly in the dryer seasons. Some ironworks drew water directly from nearby rivers or used water-powered pumps to raise water from them; others developed their own storage and supply network. The development of steam engines to drive pumping machinery, while reducing the need for industries to be located alongside rivers and streams, increased the need for flexibility and security of water supply. Storage for use in the low river flows of summer could be effected by using natural lakes but the construction of impounding reservoirs by the various industrial companies themselves became necessary as demand increased.

In the early 19th century private water companies were established but only supplied those who paid for the service. Although most of the earliest large industrial town and city water undertakers were private companies they were never financially viable. Large up-front capital expenditure meant that the cost of the water supplies to provide a return was too high for the great majority of consumers – the urban factory workers – to afford. In some cities the proportion of the population supplied with piped water actually fell in the early Victorian Period.

There was no link to drainage or sewage disposal. Rivers, streams and streets were open sewers and it was not until the connection was identified between water, waste and cholera that the need for a clean water supply and proper disposal of sewage was widely recognised. This recognition coincided with the population explosion and the rapid growth of towns and cities. This period was also dominated by the politics of improving public health and sanitary conditions. A need for a supply of clean potable water at an affordable price led to the development of private and municipal water undertakings. The critical danger signal in public health engineering, aggravated by the arrival of cholera in 1831, had been the rapidly rising death rates. Water supply and waste water and sewerage disposal needed to be brought under one local organisation for proper management and control.

An 1845 report by the Health of Towns Commission found only six out of 51 towns had a good water supply and none had adequate drainage. Thirteen were classified as 'indifferent' and the rest 'very bad'. Borough corporations had been established by the Municipal Corporation Act of 1835 but they were not using their powers to provide these services. Generally it seems that commercial provision was initially the preferred means, but this meant a lack of public accountability.

From the 1850s private bills were approved by Parliament for water supply in many Welsh towns including Cardiff, Swansea, Bangor, Denbigh and Wrexham. However, as the demand continued to grow over the next 20 to 30 years most of these towns bought out the private undertakings and took control, enabling a combined approach to both supply and waste disposal. By 1881 over 80 per cent owned their own water works, double that of 1861. By 1901 this had risen to over 90 per cent.

By the 1910s there were 2,160 water undertakings in Britain, of which 786 were local authorities. Water Boards were a mixture of small Rural District Councils, Urban District Councils and private undertakings. Most of the private undertakings would have been small local ones, particularly in 'estate' villages where the landowner would have a water supply to the big house and serve estate cottages at the same time. Some organisations such as local Boards of Health also developed water supplies. The Water Act of 1945 consolidated previous legislation and encouraged the amalgamation of water undertakers to form water boards. As a result of amalgamations, by 1963 there were just 79 water undertakings, of which 29 were private companies and 50 local authorities.

The Water Act of 1973 created 10 regional water authorities in England and Wales to take over water and sewerage responsibilities as well as water conservation, pollution control and land drainage. These were set up not on geographical boundaries but on catchment areas. For Wales there were two covering the Principality, established in April 1974. One, Welsh Water, overlaps into England in a few places and the other, Severn Trent, although mainly in England, overlaps into Wales. In addition to the 10 regional water authorities there are 14 water-only companies, one of which is in Wales, Dee Valley Water, which overlaps Welsh Water in North-East Wales.

The amalgamation of local water undertakings into larger regional ones had the advantage of a greater strategic approach to the supply with more interchange between adjacent areas and greater flexibility. Although adjacent authorities had often collaborated on enhancing supplies, having one over-arching organisation is a benefit to strategic planning. These authorities were privatised in 1989 but under regulatory control. The 10 water authorities became public limited companies and all assets were transferred to them. In the same year the National Rivers Authority was established and the regulatory framework for the water industry, OFWAT, put in place.

A Welsh example of the involvement of civil engineers in the improvement of public health is that of George Thomas Clark who was closely involved in the promotion of potable water supplies and good drainage as a public health issue. In 1843, after working with Brunel on the GWR from 1835 as one of his assistant engineers and inspectors (although his initial training had been as a surgeon), he went to India as a civil engineer. He returned to set up his own private consultancy in London and became a Public Health Inspector for the Board of Health. He reported on over 40 cities and towns in Britain. One of these was Newport in 1848. In the aftermath of a major cholera outbreak there he reported on the general sanitary state of the town and recommended the local authorities adopt the wider powers available under the Public Health Act giving powers to install main drainage and provide clean drinking water. The town council accepted his recommendations and achieved powers to that effect in 1850.

Later he became involved directly with improvements in Merthyr Tydfil. An association with the Guest family of Merthyr had developed during his time on the Taff Vale Railway. He became the resident trustee of the Dowlais Works of Merthyr Tydfil in 1855 following the death of the ironmaster Sir Josias John Guest in 1852. From this association he became involved in running the ironworks and was instrumental in converting the Dowlais works in 1865 to steelmaking using the Bessemer process, the first to do so in Wales, and later planned its move from Merthyr to Cardiff to be nearer the import point of iron ore from the company's mines in Spain.

When at Merthyr he became involved in public health issues and was deeply committed to improving the appalling living conditions in the town. He became Chairman of the Board of Guardians of Merthyr Poor Law Union and of the Local Board of Health and was closely involved in improvements to public health, including planning the provision of clean water supplies from new reservoirs in the Taf Fach valley.

The Engineering of Water Supplies

In the mid- to late 19th century potable water supplies to meet the growing demand were developed by municipal authorities, often from distant reservoirs constructed in the upland catchments. Large dams were built with a system of long-distance supply aqueducts and pipes to treatment works nearer the point of use.

As an alternative to a reservoir with an aqueduct, a regulating reservoir can be constructed. The river below the dam can then be used as the aqueduct, carrying water many miles downstream to extraction points at towns lower down. Water can be stored during the winter and river flows can be maintained during dry periods by releasing water to ensure sufficient supplies at the take-off points downstream. As long ago as the early 1800s Telford raised the level of Lake Bala, constructing a dam and sluices to maintain dry weather flows in the River Dee, ensuring adequate water in the summer at **Horseshoe Falls Weir (N38)** for the Llangollen Canal.

Dams can be of several different forms of construction depending on the topography and geology of the area. They can be constructed to increase the water storage capacity of existing lakes or to create a new lake in a geologically suitable valley. In some cases it is possible to use a lake as a freshwater source without storage enhancement. Embankment dams are constructed to resist water pressure with their weight. They have a vertical impermeable central clay core to prevent seepage through the dam, which has a flat downstream slope, usually grassed, and a steeper upstream slope, usually stone-pitched to resist wave action. They can be of compacted earth or rockfill. This form is particularly suitable for wide flat valleys. Rockfill types can have an impermeable upstream face rather than a core, usually of asphaltic concrete. Gravity dams are constructed of masonry or concrete, also resisting water pressure with their weight, but because of the greater mass of the material the downstream and upstream slopes can be quite steep. This form is suitable for narrower valleys with sound rock for a foundation. A development of this type is the buttress dam with the downstream face supported by buttresses, reducing the amount of material necessary to resist the water pressure. Arch dams in masonry or concrete can be used where the rock formations at the sides of the valley are able to support the thrust from the arch form. These are more economical of material and capable of being built to greater heights. Sound rock is also needed for the foundations.

Spillways are necessary for all dams to release water when the reservoirs are full. For embankment dams these are usually side channels formed in masonry along the side of the dam with an inlet point just upstream of it; some are stepped to reduce the speed of the water and avoid erosion at the base of the dam. A stilling pond is often provided at the base to reduce the force of the water. Some gravity and arch dams have a lowered section of the crest, which allows water to spill over the top and down the face. A third type is a shaft spillway, where water flows into an overflow pipe upstream of the dam and then through a tunnel to downstream of the dam. Most dams have a sluice arrangement built in to release water when the reservoir level is below the spillway so as to maintain flows in the river or stream below the dam site. Valve towers are built into the dam or nearby for the stored water to be taken into the distribution pipe network or aqueduct for transmission to the treatment plants, which can be many miles away.

Central and North Wales became the main water catchment areas for the English cities of Birmingham and Liverpool, with pipelines and aqueducts many miles long across the border. Liverpool had started investigating possible additional sources of supply in 1866 and in 1880 selected the Vyrnwy Valley in North Wales, building the first large reservoir in Wales and the first masonry dam in Britain (**N34**), with a 65-mile-long aqueduct to Liverpool.

Birmingham selected the Elan Valley (**M12**) in Mid Wales. Work started on the first phase in 1893 and was completed in 1904. Four reservoirs were created along the valley, the uppermost being Craig Goch, then Penygarreg, Garreg Ddu and Caban Coch, and a 73-mile aqueduct was built. The water flows under gravity to Frankley Treatment Works near Birmingham at about two mph, taking a day and a half to complete the journey. Claerwen Dam was added to the system in 1952.

The Alwen Dam (SH 955525), one of the tallest of the early concrete dams, was constructed between 1909 and 1921 in Denbighshire by the Corporation of Birkenhead to supply piped water to the Wirral and Liverpool. It now supplies five million gallons a day to North Wales. Llyn Brenig (SH 970545) was formed between 1973 and 1976

69 *Claerwen Dam.*

70 *Penygarreg Dam.*

71 *Clywedog Reservoir.*

72, 73 *The 20hp beam engine installed at Ely Wells in 1851 with 12ft-diameter flywheel. Now in the Waterworks Museum at Hereford, on loan from the National Museums of Wales. (© Water Museum Hereford)*

in a neighbouring valley to regulate the water supply to the River Dee so that water could be extracted further downriver. It is one of the largest areas of inland water in Wales.

In 1955 the **Usk Dam and Reservoir (S33)** was built to provide an additional supply to Swansea. It is also used as a regulating reservoir for the River Usk. The Clywedog Reservoir (SN 912870), built in 1967, acts as a regulator for flows in the River Severn controlling winter flows to minimise flooding downriver and releasing water in summer to maintain river flows for abstraction by the many English towns along its length. **Llyn Brianne Dam (M15)** was built in 1972 as a regulating reservoir on the Towy River for West Glamorgan Water Board, the take-off point being 40 miles downstream. This dam is the highest in the UK. A hydroelectric power station was added in 1997.

The development of town water supply can be illustrated by the example of the city of Cardiff. The main water supply in around 1800 for a population of around 2,000 was from public wells in the centre of the town and from nearby streams. By 1840 the town population had grown to about 10,000 and these facilities were becoming inadequate. In 1850 a private company obtained powers to develop a water supply and built a pumping station at Ely, three miles from the town, to abstract water from the River Ely. This was roughly filtered and raised to a service reservoir on higher ground nearby, from where it could be distributed by gravity through a series of mains laid through the town. Abstraction was limited to three million gallons a day. The first storage reservoir was built by this private company at Lisvane in the north of the town in 1865 (60 million gallons capacity), gravity fed from local streams and then piped to a treatment plant and distribution pipework.

In 1879 Cardiff Corporation bought out the private company. By 1885 the supply was again becoming inadequate and an expansion was planned. In 1892 Cantref Reservoir (322 million gallons capacity) was built near the head of the Taff Fawr River, 32 miles away, and the water piped to a new reservoir at Llanishen built alongside the original one at Lisvane. The scheme was capable of supplying 12 million gallons a day. Plans allowed for the construction of additional dams and in 1897 Beacons Reservoir (345 million gallons) was added. Later in 1926 Llwyn On Reservoir (1,260 million gallons) provided additional storage and a total capacity of 24 million gallons a day (SN 987182, ST 186818). The relative populations of the town over this period are – 1801: just under 2,000; 1851: 18,000; 1871: 60,000; 1901: 160,000.

The three dams are all earth embankments with a clay core, stone-pitched on the upstream side and grassed on the downstream face. Each has coursed rock-faced valve houses to take off the water to the pipe network, and a stone-paved spillway with dressed stone revetments. The reservoir at Llanishen is built on flat ground as a complete encircling reservoir above ground level to the same construction. The dam here is 0.9 miles long and has a capacity of 317 million gallons with a water area of 60 acres. A new treatment plant and covered reservoir was built in the mid-20th century a few miles north-west of Llanishen reservoir, which is now used solely as a recreational facility.

Demand for water continued to grow and in the mid-20th century further reservoirs were built, this time jointly with other towns, by the new regional water authority, which was empowered to plan a comprehensive supply system for the whole area and the supply to Cardiff was enhanced once again. The original wells at Ely continued to supply about one million gallons a day until 1926. Water was used at the nearby paper mill and later piped to the power station at Aberthaw.

74 *Ynys y fro Reservoir, Newport.*

In the 19th century other towns and cities also developed their systems. Ynys y Fro Reservoir (ST 285889) constructed in 1848 was the first dam built to supply Newport (after the report by G.T. Clark following the cholera outbreaks). The engineer was James Simpson, later seventh president of the Institution of Civil Engineers. The reservoir had a capacity of 71 million gallons. It was extended in 1872 by a new dam at its upper end, adding an additional 33 million gallon capacity. The population of Newport had grown from about 750 in 1790 to 1,135 in 1835 but had reached 19,000 by 1845.

Nant Hir Reservoir (S31) was built in 1875 to supply Aberdare and District. Four reservoirs were constructed in the Taf Fechan Valley to supply Merthyr Tydfil UDC. The first was Pentwyn (or Dolygaer) (SO 055145) in 1863 by the Merthyr Board of Health, then Lower Neuadd in 1884 by Merthyr Tydfil Corporation, Upper Neuadd (**S32**) in 1902 and Pontsticill in 1927 (SO 058119).

In 1939 Talybont Reservoir (SO 104206), the largest in the Brecon Beacons, was completed to supply additional water to Newport. In 2006 the first community-owned hydroelectric station was commissioned at Talybont. There had been an old turbine house built to supply power to the treatment works there which was closed when the National Grid reached the valley. The local community have recently provided a new turbine and refurbished the old turbine house as a 'green' initiative and power is now supplied to the National Grid as well as locally.

Llandegfedd Reservoir (ST 330995) was added to the network in 1979, supplying up to 20 million gallons a day to Cardiff, Newport, Abertillery and Pontypool.

Wrexham provides another example of the development and gradual expansion of water supply. In the 1850s Wrexham had a population of about 7,500, and in 1863 a group of townspeople established the Wrexham Water Company, whose works to pipe water from the Pentrebychan Stream were completed in 1867. These comprised an abstraction reservoir, a storage reservoir and sand filters with a pipeline to the town, parts of which are still in use. A further impounding reservoir was

75 *Valve House, Talybont Reservoir.*

76 *Legacy Tower, Wrexham. (© Eirian Evans)*

constructed in 1878 as the town's population had grown to about ten thousand. By 1901 the town's population had grown to nearly 15,000 and a further reservoir was completed in 1904, by which time the company was also supplying parts of Cheshire and Flintshire. Additional powers were obtained to abstract water in 1921 and the water tower at Legacy (SJ 295483) was constructed in 1934 for water treatment. In 1933 further supplies were pumped from the lower workings of the Minerva Mines and in 1951 powers were obtained to abstract six million gallons of water a day from the Dee, increased to nine million gallons in 1975. The company converted from a statutory water company to a plc in 1974 and merged with the Chester Water Company in 1997 to form Dee Valley Water. Wrexham alone has a population of over 45,000 now.

There are relatively few water towers in Wales as the hilly country and natural heads of water are usually sufficient to provide adequate pressure. Often supply reservoirs are constructed on high ground as an alternative to a tower, providing storage to allow for fluctuations in demand and to maintain adequate pressure in the distribution pipework. One listed redundant water tower in Penylan in Cardiff has been converted for residential use (ST 194792).

Many of the industrial complexes required their own water tanks to supply the factories and works. These were often of modular steel construction, steel panels forming a rectangular tank supported on structural steelwork, often termed 'Braithwaite' tanks after one of the manufacturers, still based in South Wales, who patented their system in 1901. The tank can be designed to a range of capacities up to 15,000 cubic metres by adding more panels.

The railway network also had water towers at regular intervals to provide a water supply for steam locomotives. Most have been demolished and few remain on the main network. There are some small examples on the narrow-gauge steam railways. There is a redundant reinforced concrete Grade-II listed railway tower at the west end of Cardiff Central Station (ST 183759).

One notable water tower is on high ground near Neyland in West Wales (SM 950069). Here the water tank is one of the first examples in the UK of a prestressed concrete water tank. (The earliest example is at Meare in Somerset, built about 1953.)

Below, from left to right:

77 *Penylan, Cardiff.*

78 *Typical 'Braithwaite' tank.*

79 *Cardiff Central Station.*

80 *Water tower at Neyland.*

WASTE WATER TREATMENT

The treatment of waste water and sewage is possibly one of the less visible aspects of civil engineering. Here the work of the civil and municipal engineer is a mainly unseen but nevertheless vital part of the infrastructure essential for modern living. Most of the engineering is below ground and often the only visible signs are the pumping stations provided to maintain the flows in the sewers. Originally using steam-powered pumps, these have been converted to electricity or replaced with new buildings as the systems have expanded and been extended. Some of the older disused pumping station buildings can still be found, some converted to other uses.

Although as long ago as Roman times drainage systems had been engineered to collect waste water, disposal was usually direct to the nearest stream or water-course. However, once the link between foul water and diseases such as cholera was accepted in the mid-19th century the need for controlled disposal to prevent contamination of drinking water wells and springs was recognised. Civil and municipal engineers designed and built collection systems, pipework and pumping stations for the collection of waste water and chemical effluents.

Edwin Chadwick (an improver-bureaucrat of the early Victorian period), after his *Report on the Sanitary Conditions of the Labouring Population of Great Britain in 1842*, emphasised systematic drainage as a more urgent priority than better house building which he and others previously urged. His lasting legacy in engineering terms is his recommendation for the construction of sewers with adequate water supply to flush them clean. One of Chadwick's many contributions to the development of public health engineering was to facilitate legislation in 1847 setting out standard waterworks clauses which municipalities could include to ease and hasten the passage of private Bills for their own corporate purposes and powers. The success of this legislation can be measured by the fact that within 10 years there were 78 municipal water undertakings compared with only 11 in the decade before Chadwick's report, and numerous companies as well.

Up to the 18th century there had been little in the way of foul sewers, as there had not been much indoor sanitation. Such drains and sewers as there had been in the towns were almost entirely for the clearance of rainwater. The expanding towns needed larger sewerage systems. New built-up areas increased surface drainage flows and the system generally developed as separate foul sewers at the rear of dwellings and surface water drains at the front. Occasionally combined sewers were built to assist in maintaining a self-cleaning flow in dry weather. It soon became evident that sewage

81 *A 48in storm water drain and river outfall.*

disposal would be at least as much of a problem as water supply had been before the larger towns began to master it. Various proposals, initiatives and pieces of legislation were put forward to deal with this second major area of urban water organisation, leading onto a series of four Royal and other Commissions between 1865 and 1874. They recognised the force of some of the arguments about municipal boundaries not being always well suited to separate water supply and sewage disposal organisation. Arising from this series of studies the Public Health Act of 1875 and the River Pollution Prevention Act of 1876 were enacted.

Essentially these laws aimed to do two things: to make it an offence to discharge or dump sewage and industrial or mining wastes into rivers, and to appoint the municipal councils as the enforcement agency for these new restraints. However, in practice the polluters would have a good defence if they had used the best practicable means of rendering harmless the wastes and effluents being discharged.

In the 1930s there was a run of further legislation, on fisheries in 1923, land drainage in 1930, on public health and effluent disposal in 1936-7 and the Water Act of 1945 updated and strengthened the Victorian water supply Acts of 1847 and 1875.

From the mid-19th century, therefore, sewers were built to collect waste water. These systems were constructed as a series of gravity systems. Pumping stations were built at intervals to connect drainage networks and to raise the water to a higher level for it to fall again by gravity or under pressure to the next pumping station, thus enabling extensive networks of drains and sewers to be provided, with a terminal pumping station for the final discharge to treatment works, watercourse or sea outfall. Perhaps the most well-known example is the work of Joseph Bazalgette in London in the 1860s and '70s where a vast new sewer network was built to intercept existing short sewers and watercourses discharging into the Thames. The construction work transformed the banks of the River Thames in Central London with new river walls and highways.

Separate storm water systems from paved areas and roads may be discharged directly to watercourses. Industrial waste water has to be treated by the company to a standard acceptable to the water undertaker before discharge into their network or water courses.

Sewage treatment works normally consist of a screen to filter out large waste matter, paper, plastic, stones etc.; a grit tank where small stones and grit settle out; a sedimentation tank where waste material would settle out and be collected as sewage sludge for disposal; and a filter bed to improve the quality of the water being discharged to water courses, rivers or out to sea. In the early days treatment was fairly rudimentary.

In the 20th century the design of treatment works developed further, with additional treatment in digestion tanks where bacterial activity broke down organic matter in the effluent and filter beds improved the effluent quality. Effluent was still discharged to rivers or by long sea outfalls. A by–product of this treatment is methane gas, which can be used to generate electricity to power the works.

82 *Cardiff pumping station, built c.1900 (now an antiques sale-room).*

In the latter part of the century there have been major developments in the treatment of foul and industrial waste water. Very large treatment plants have been built, with associated intercepting and collecting sewers bringing water large distances to these plants and improved technology to purify the effluent. Significant Welsh examples include the new Waste Treatment Works at Cardiff and Swansea. The final effluent discharged to rivers or sea is now in most places of a high quality and our rivers and seas much cleaner places.

The treatment of storm water collected in a combined surface water run-off and foul water sewer has also improved. While this was not a problem in normal conditions, in severe storms the treatment works would be overwhelmed by the volume of water entering the system. At these times the combined effluent would be discharged directly, bypassing the treatment works on the principle that the large quantity of rainwater would dilute the foul water to an acceptable level. This is no longer an accepted method and special holding tanks and larger diameter sewers are provided to store the water until it can be treated.

POWER

The construction of weirs, sluices and structures for generation of water power and the design and construction of all the structures, storage and transmission facilities for the various forms of power generation required the skills of the civil engineer from early times, even if the profession was not recognised as a distinct one until the early 19th century (see p. viii). This involvement continued even after the formation of various more specialised institutions such as the Institution of Mechanical Engineers (1847), the Institution of Gas Engineers (1863) and the Society of Telegraph Engineers, later the Institution of Electrical Engineers (1871).

Water Power

Sources of power in medieval Wales would have been wind and water. There are few windmills remaining in Wales; in fact only one has been restored and is in operation at Melin Llynon, Llandeusant, Anglesey (SH 340853), but many small water-powered mills of various ages survive. Around twenty, mostly corn mills, have been restored to working condition. A recent estimate suggested that there were over 3,000 water-power sites in Wales in the 19th century. Many of the early industries would have used water wheels to provide the power for machinery and to pump water. This needed the design and building of weirs, leats and water channels to power wheels for mills, ironworks and other industrial sites, notably coal mining in South Wales, metal mining in Mid Wales and slate quarrying in North Wales. Some of the mining leats, especially in Mid Wales, were impressively engineered and of a prodigious length.

A feature of the system is often a mill pond, fed from a weir at a suitable location upstream to provide a reserve of water at times of low flow with an artificial channel to the wheel. The wheel is driven by the flow either to the top of the wheel (overshot) or the bottom (undershot), the former using the weight of the water as well as its force to turn the wheel. The wheel then drives a large vertical gear wheel inside the building which in turn drives other gears and belt drives in the mill or workshops, powering the machinery.

Several examples of industrial water wheels still exist throughout Wales, the largest being the de Winton Wheel (SH 585602), located in the workshops of the original Dinorwig slate quarry, now the National Slate Museum in Llanberis, North Wales. Here it is possible to see how the machinery in the various slate workshops of the Dinorwig Slate Quarry was belt-driven from the water wheel.

83 *De Winton Water Wheel, Llanberis. (© National Slate Museum)*

At Llywernog Lead Mining Musuem, Mid Wales (SN 735808) a working water wheel driving typical mining machinery can be seen, and the Museum of the Welsh Woollen Industry at Drefach, Felindre, near Llandyssil, West Wales (SN 354386), retains a water wheel to drive the looms and weaving machines there. Cyfarthfa Ironworks in South Wales had one of the largest wheels in Wales, using water from the Taff brought by a long aqueduct from higher upriver to provide power for the works.

Tidal power is not a new concept and extensive use was made of this energy source in earlier times, although usually on small-scale installations. In West Wales there is a mill dating from the 15th century that used tidal power to drive its water wheel. **Carew Tide Mill (S43)** sits on a dam built across the lower tidal section of the Carew River. As the tide comes in, the sluices are opened and the dam filled. As the tide goes out the water is retained and then fed to a waterwheel in the mill to generate the power for grinding corn. The mill has recently been refurbished.

Hydraulic power was common particularly in docks and harbours. Accumulator towers located around the dock area were used to pressurise hydraulic pipework and this pressure was used to operate dock gates, sluices, bridges and even cranes. Heavy weights were raised to the top of the tower on a piston to pressurise the system, and pressures up to 750psi could be achieved. Prior to this development the only pressure available was by gravity, limiting the effectiveness of the system. Hydraulic power was in use in the South Wales ports as recently as the 1950s, when it was phased out and replaced by electric power.

84 *All of the hydraulic towers have disappeared except one by Brunel at Briton Ferry, which has recently been restored although the mechanisms inside have gone.*

Steam

The development by James Watt in 1763 of the original atmospheric beam engine designed by Thomas Newcomen of Devon, adding a separate condenser to give it greater efficiency, enabled industry to develop the use of steam power, losing its ties with water power and the need to locate near flowing water. Trevithick's improvements in the late 18th century using high-pressure steam were a further major advance. The Cornish beam engine improved the Boulton & Watt design with plunger pumps better matched to the duty and was used from about 1810. Static steam engines became commonplace, driving machinery, pumps, etc. Improved, higher-powered, pumping engines enabled the mining of minerals at deeper levels and at the same time created a greater demand for coal. The introduction of coal-fired boilers for locomotives and ships was a further boost to the mining industry as well as a major transport development.

Steam pumping engines were built for water supply and for foul drainage discharges as well as dewatering mines. In the 19th century the size of these machines grew rapidly. Colliery winding gear was driven by steam power as well as being used elsewhere in the above-ground workshops.

There is a major pumping station at Sudbrook in Monmouthshire (ST 507874) which pumps between 12 and 20 million gallons of water daily from the Great Spring under the Severn Rail Tunnel. Originally steam-driven, the pumps have now been electrified. There were six huge Cornish beam engines operating until 1961. The beams were 30ft long and weighed 20 tons. They were manufactured by Harvey & Co. of Hayle, Cornwall.

85 *One of the beams of the original engines is on display at Swansea Museum.*

ELECTRICITY GENERATION

Although the principles of the electric motor and transformer were discovered in the 1830s it was not until the late 1800s that the practical use of electric power became widespread. The key invention was the incandescent lamp invented by Swan and Edison, which replaced the use of impractical carbon arc lamps. The main initial demand was for street lighting, replacing gas; the first public use was in Godalming in 1881 where the electricity was generated by water power. Later, coal, oil and gas provided the fuel source for steam turbines to generate electricity, and in the late 20th century nuclear fuel has grown in importance. Where the geology and geography were suitable, hydroelectric generation has been used. The replacement of the older gas lighting with this new, cleaner, more flexible alternative was widely welcomed. It was soon extended to industrial and domestic use. Power companies were formed to manufacture and supply electricity locally in towns and cities.

By 1915 there were over 600 electricity undertakers in the country and it was proposed that District boards should be set up to take over power generation and distribution. In 1925 the Central Generating Board was established to operate the National Grid system, interconnecting the biggest and most efficient power stations. The industry was nationalised in 1948 with the creation of the Central Electricity Generating Board, which itself was denationalised in 1990. By then a range of power sources were in use to generate electricity.

The first uses of electricity in Wales can be traced to South Wales. Cardiff in 1893 had a small steam-generated direct current (DC) supply for arc lighting which was introduced for street lighting in the town. This superseded gas lighting introduced in 1821. Another early scheme for street lighting was opened at Nantymoel, Ogmore Vale, in 1891. Use in Cardiff expanded rapidly with new small power stations producing alternating current (AC) supplies in 1894. The use of electricity grew throughout Wales, including Pontypool 1893, Newport 1895, Swansea 1900, and Merthyr 1901, and further small generating stations were built. The tram system in Cardiff was electrified in 1900. Wrexham's horse-drawn trams were converted to electric in 1907 after electricity generation there from 1900.

In Wales we had one of the early water-powered electricity generating stations. This was at Monmouth opened in 1899, initially for street lighting. The station used water diverted from the River Monnow at Osbaston Weir (SO 502138) and it remained in use until 1950. In case of low river flows it had a standby steam boiler, later changed to a diesel engine. The station has now been demolished. For the latter part of its life it was used as an engineering workshop. The weir is still in existence but the leat to the old power station has been filled and the land regraded.

Apart from Monmouth there are other hydroelectric schemes in Wales. In 1928 four dams were built in North Wales to create Trawsfynydd Lake to feed a hydroelectric power station at Maentwrog (SH 673377). A nuclear power station opened at Trawsfynydd in

1968 (closed in 1991) used the lake for cooling water, but restricted the hydroelectric station to a maximum change in water level of five feet. In 1961 a hydroelectric scheme was built in the Rheidol Valley (**M4**). Three high-level dams were constructed, feeding a pipeline to the power station. The station capacity was increased in 1996. Llyn Brianne had a hydroelectric power station added in 1997. Talybont Reservoir also now has a small community-owned hydroelectric station, opened in 2006. The Centre for Alternative Technology near Machynlleth has several examples of the use of water power for energy production.

The development of hydroelectric schemes now extends to the use of estuarial barrages and tidal power for electricity generation, which together with wave power will give the current and next generation of civil engineers more challenges for their ingenuity.

In modern times the use of wind power to generate electricity is developing rapidly and the early wind farms may eventually become sites of historic interest. The massive towers and turbines are mainly mechanical and electrical machines but the civil engineer provides access to the remote areas best suited to wind power and the foundations on which the towers sit.

Electricity Storage

The construction of commercial nuclear power stations from 1956 onwards, which need to be constantly generating electricity and producing a base load, led to the need for a system which could store the surplus energy produced overnight for re-use during peak daytime periods. This resulted in the development of pumped storage schemes where water is pumped up to a high-level lake using off-peak electricity and released through turbines to a lower lake to generate hydroelectricity at times of peak demand.

In Wales we have one of the earliest of these at Ffestiniog (**N26**), opened in 1963, and one of the world's largest at Dinorwig, Llanberis (**N27**), opened in 1984. Both use natural lakes for high- and low-level reservoirs, although each lake has had its water capacity enhanced for power generation. Their construction presented major challenges for the civil engineer, not the least being the huge scale of the construction work

GAS

The role of the civil engineer in the gas industry can be found mainly in the infrastructure necessary for its transmission, distribution and storage. In conjunction with the other engineering disciplines, mechanical, chemical and gas engineering, the design and construction of these facilities would have involved many early civil engineers as support to the main production facilities. Generally the civil engineer would have worked as support to the gas engineer who would lead the design and construction teams. In many cases the gas engineer would be a member of both Institutions.

The production of gas from the distillation of coal was known as early as the 16th century but it was not until the beginning of the 19th century that its manufacture and use commercially became practicable. William Murdock, a Scot working as a site engineer in charge of Boulton & Watt's steam engines in Cornwall, first used coal gas for lighting in his house in Redruth in 1792 and by 1798 had invented the equipment to store the gas, which was originally a by-product from the production of coke from coal. Experiments in the use of gas for lighting were also being carried out in France, and Gregory Watt, son of James, knew of these and backed Murdock's work. Commercial gas plants for large mills were produced by Boulton and Watt from 1804, the first being installed by Murdock's pupil, Samuel Clegg, for lighting a cotton mill at Sowerby Bridge near Halifax in 1805, followed by Murdock's installation in a cotton mill in Manchester in 1806, and by 1814 over 30 small mill gas works had come into existence.

By this time coal gas had been used to light streets in Westminster, supplied from the first city gas works built in 1812/13. This had developed from independent demonstrations of gas lighting by Frederick Winsor in London in 1805/7, with an alternative system of supplying gas from a central production plant. Cardiff had gas lighting in 1821 and Wrexham in 1827.

By 1830 almost every city and large town in the country had a gas works, mainly built by private enterprise. This required the construction of gas holders and gas distribution systems to store the gas and maintain pressure. The telescopic gas holder was developed in 1824. Lighting for the first 75 years was by open flame as the incandescent mantle was not invented until 1885.

Following the invention of the Bunsen burner in 1855 the use of gas developed for cooking and heating, a significant improvement on coal. The first practicable gas fires followed over the next 20 years. In the towns, as well as street lighting, most public buildings and large stores had a gas supply, but it was not until the late 19th century that most workers' homes had a supply after the introduction of the pre-payment gas meter.

Winsor's original gas company, the Gas Light and Coke Company, continued until the gas industry was nationalised in 1949. The industry was privatised again in 1986. In 1994 the privatised British Gas split into two companies with one dedicated to the transmission of gas via a national network.

Natural gas can be found associated with oil or as an independent resource. It can also be manufactured from oil products. The first shipment of liquefied natural gas from Algeria arrived in Britain in 1964, but in 1965 gas was discovered in the North Sea and between 1968 and 1976 Britain changed from the use of coal-derived town gas to natural gas, with major construction work taking place around the country for storing and distributing gas from the gas fields in the North Sea. Over 275,000km of pipelines and gas mains now exist with compressor stations to maintain pressure and new liquid and underground storage facilities for the gas. Additional installations to enable the further import of liquefied natural gas by sea and new connecting pipelines bringing gas from Europe were built recently as North Sea gas supplies are depleted.

In Wales in particular new jetties and import and storage facilities have been constructed in Milford Haven, with new and upgraded major pipelines linking to the existing transmission network. All this has involved the expertise of the civil engineer in design and construction.

Land drainage and flood defence are generally understood to include the alleviation or control of flooding in urban or agricultural land, whether by fresh or salt water, including the improvement and maintenance of natural and man-made channels. Reclamation, particularly of tidal areas, will often also include protection works for the reclaimed land. Land reclamation can also arise as a consequence of river works for the construction of unrelated other facilities such as docks and harbours.

A particular example can be found in Cardiff when in the late 1840s Brunel straightened the River Taff in Cardiff to facilitate the construction of Cardiff station on the South Wales Railway. This cut off a large bend of the river which had originally been used as a riverside wharf. The land remained undeveloped for many years but eventually as the town expanded housing was built near the station and the wharf area redeveloped. The remaining land became a sports ground, world-renowned as Cardiff Arms Park and now the Millennium Stadium, home of Welsh Rugby.

LAND DRAINAGE AND RECLAMATION

Statutory management dates back to Tudor times with Acts such as the Statute of Sewers in 1531. Various drainage Acts have been promulgated, culminating in the Land Drainage Act of 1930 which set up Catchment Boards and Internal Drainage Boards in England and Wales, the first of these to have oversight over the main rivers. The Boards were to be funded by other authorities including county councils. The Rivers Boards Act of 1948 brought together the responsibility for drainage, fisheries and pollution under single authorities and in 1951 in Wales six River Boards came into being, Wye, Usk, Glamorgan, South-West Wales, Gwynedd and the Dee & Clwyd River Boards. In 1964, following the Water Act of 1963, the River Boards became River Authorities responsible for flood prevention and land drainage along the main rivers, together with conservation and augmentation of water resources, pollution control, water quality and fishing. There are 14 Drainage Boards in Wales, responsible for the drainage of low-lying land, almost all in coastal and estuarial areas.

The reclamation of low-lying salt marsh and tidal flats was seen as an economic investment with high returns as the land was rich in alluvial soils and offered high crop yields. The land could be worth as much as twice that of dryer arable land. As a consequence there were many large and small reclamation schemes around the coast.

The earliest examples of land reclamation date back to Roman times. In North Wales the Romans straightened the River Dee for improved sea access to Chester and reclaimed land at the same time. Since then there have been successive additional works to form or improve harbours at Flint and Connah's Quay. In the 1730s there were major works to construct an eight-mile-long improved channel through low-lying reclaimed land from Chester to Connah's Quay (**N45**) because of siltation in the existing channel to Chester. In medieval times access improvements to the harbour, town quays and castle at Rhuddlan included land reclamation (**N33.1**) and there are few low-lying areas around the whole Welsh coast that have not been reclaimed from salt marsh or mud flats. As well as at the Dee land has been reclaimed at Abergele, Towyn, Prestatyn and Rhosemor. Malltraeth Marsh in Anglesey was the subject of several failed attempts in the late 18th century. Eventually in 1788-9 a 'cob' or dyke over half a mile long, designed by James Golborne, who had worked for his uncle John on the Dee reclamation work, and built by Pinkerton & Dyson, was completed across the bay allowing land reclamation behind it.

86, 87 *Brunel's river diversion at Cardiff. Left – lower end of diversion and rail bridge looking north; the station is out of picture on right. Brunel's bridge (centre) has been widened on both north and south sides. Below – upper part of diversion., Millennium Stadium, Cardiff Arms Park, is on part of the reclaimed land on right.*

88 *View of the Levels at Wentlooge.*

89 *Typical drainage reen and sluice.*

In West Wales 280 acres were reclaimed when Goodwick Brook was canalised in late medieval times. In South Wales there are extensive land drainage and reclamation works. The Gwent Levels east and west of Newport were reclaimed from the Severn Estuary, or protected from inundation, by sea walls, some parts of which may have been built by the occupying Roman legions based at Caerleon, north of Newport. Some of the higher ground is known to have been occupied in the Bronze Age but no evidence has been found of contemporary sea defence works. Many square miles were drained or reclaimed in medieval times and a comprehensive system of drainage ditches (known locally as 'reens') and sluices has been provided, possibly partly as protection of low-lying land as sea levels rose during the post-Roman period. The area is now managed by the Caldicot and Wentlooge Levels Internal Drainage Board. Near Caldicot the medieval field patterns can still be traced.

The most notable example in North-West Wales is the reclamation of large areas of the Glaslyn Estuary in the early 1800s. William Madocks, of London, owned a small estate in Merioneth and in 1798 purchased some small farms on the north-west of the estuary at Penmorfa. These were each side of an inlet off the estuary and in 1800 he had an embankment constructed to reclaim the inlet, reclaiming over 1,000 acres, and developed the new town of Tremadog. He then sponsored the construction of a large embankment across the estuary itself, **Porthmadog Cob (N23)**, built between 1811 and 1815. It allowed the reclamation of more land and the construction of the town and harbour of Porthmadog. This also provided a new tolled road crossing of the estuary and in 1836 rail access for the new Ffestiniog Railway to the harbour.

COASTAL PROTECTION

Reclamation works lead to coastal defence construction to ensure the land reclaimed does not suffer further inundation. The construction of sea defences can be a project in its own right but can often be as a consequence of other construction. Stephenson's Chester & Holyhead Railway across large lengths along the North Wales Coast

90 *The construction of the new A55 also encroached on the foreshore at Colwyn Bay. Here in the 1980s substantial sea defences were provided using rock armouring and purpose-built pre-cast concrete units. (© NH)*

91 *Rhos-on-Sea breakwater. (© NH)*

between Connah's Quay and Llanfairfechan was built on low-lying and reclaimed land and required sea defences (**N33**) to ensure its stability. Further works were necessary on Anglesey at Malltraeth Marsh and Holyhead.

At Penmaenmawr the new A55 was built in 1989 over the existing beach and promenade and a new promenade was constructed seaward to replace the buried one and provide protection to the road. Additional sea defences were necessary to the east. Spoil from the construction of the immersed tube tunnel at Conwy was used to reclaim a large area of the Conwy Estuary upriver and create a new wetland habitat.

At Rhos-on-Sea an offshore rock-fill breakwater was built in 1983 to reduce the force of north-easterly storms and prevent flooding of businesses and properties on the sea front (SH 845807). This was only the second of its type in the UK, the first being Wallasey. Originally designed by Colwyn Borough Council to prevent flooding of the business district of Rhos-on-Sea, the breakwater was to have been connected to the shore by an arm along which construction materials could be transported from quarries six kilometres to the east of Colwyn Bay. However, tenders exceeded the available funds and an alternative proposal to supply the material by sea from quarries on Anglesey was accepted. Construction began in April 1982 and was completed in March 1983. The rockfill breakwater has a minimum formation level of four metres below Ordnance Datum and the crest level is 6.5 metres above O.D. The structure is founded on stiff clay below the beach material. The maximum weight of individual armour stones is about 5.5 tons and the crest length is 208 metres. The breakwater also protects the sea walls south of Rhos-on-Sea by reducing the height of the waves, allowing accretion of beach material and affording greater protection to the walls.

GAZETTEER

This section of the book is a descriptive list of civil engineering works, not only many of those referred to in previous chapters but other sites which further illustrate the contribution of the civil engineer to mankind's progress over the last four centuries.

The sites are arranged by geographically for North, Mid and South Wales. Maps show the approximate position of each site. For each site the following information is given:

- The national grid reference;
- A brief description of the work;
- If necessary the accessibility of the site and any access charges if appropriate;
- The HEW number for those on the PHEW database.

INTRODUCTION

Sites have been grouped into North Wales, Mid Wales, and South Wales. The sites have been listed in an order and grouping broadly based on county boundaries within these subdivisions, but in places topographical boundaries have been used where groups of sites relate more logically to each other. Some sites such as roads and railways cross several county boundaries.

Anyone visiting these sites may well have an interest in the general industrial history of the area. Information on important historical locations can be obtained from Local Authority information centres and tourist offices to supplement the sites in this gazetteer. Also recommended is the European Route of Industrial Heritage (information at www.erih.net) which provides a framework for visiting historic industrial sites based around an anchor site. In Wales anchor sites are Big Pit Blaenavon for South-East Wales; Rhondda Heritage Park for Central South Wales; the National Waterfront Museum Swansea for South-West Wales and the National Slate Museum Llanberis and Amlwch for North Wales.

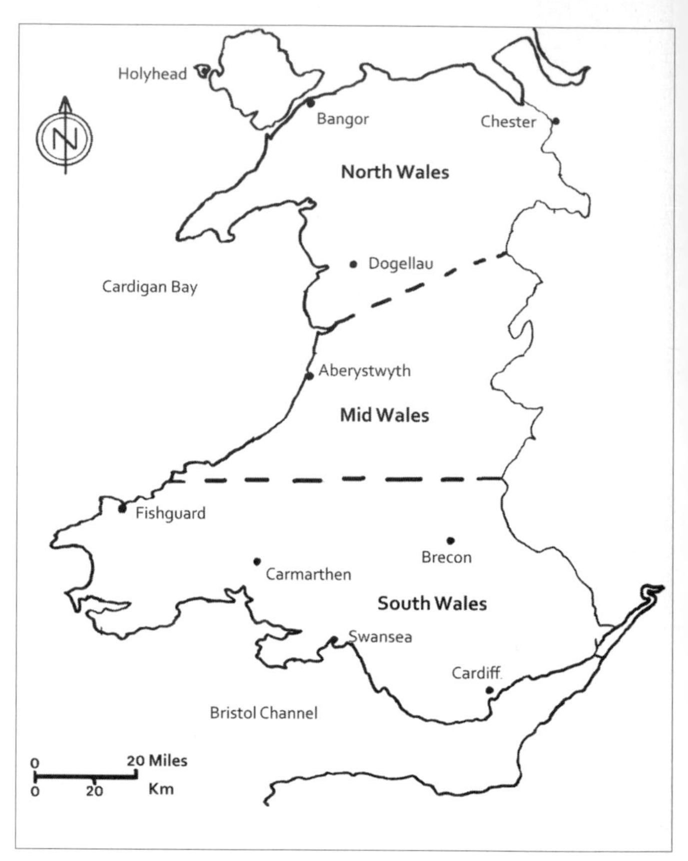

Wales map.

THE WELSH COUNTIES

Historically from the Middle Ages there were 13 Welsh counties: Anglesey; Brecknockshire; Caernarfonshire; Carmarthenshire; Cardiganshire; Denbighshire; Flintshire; Glamorganshire; Merionethshire; Monmouthshire; Montgomeryshire, Pembrokeshire and Radnorshire. In 1974 these were abolished and replaced with eight new counties: Clwyd; Dyfed; Gwynedd; Powys; Gwent; Mid Glamorgan; South Glamorgan and West Glamorgan. A further reorganisation in 1996 combined the counties with the district councils to form a single tier of 22 new Unitary Authorities. They are Blaenau Gwent CBC; Bridgend CBC; Caerphilly CBC; Cardiff CC; Carmarthenshire CC; Ceredigion CC; Conwy CBC; Denbighshire CC; Flintshire CC; Gwynedd C; Isle of Anglesey CC; Merthyr Tydfil CBC; Monmouthshire CC; Neath Port Talbot CBC; Newport CBC; Pembrokeshire CC; Powys CC; Rhondda Cynon Taf CBC; Swansea CC; Torfaen CBC; Vale of Glamorgan CBC; Wrexham CBC. Of these only Pembrokeshire, Carmarthenshire, Ceredigion (Cardiganshire) and Anglesey are similar to the pre-1974 counties. Although 11 took the title of county borough all have equal status. In the gazetteer all have been termed Unitary Authorities (UA) to avoid confusion with the earlier counties.

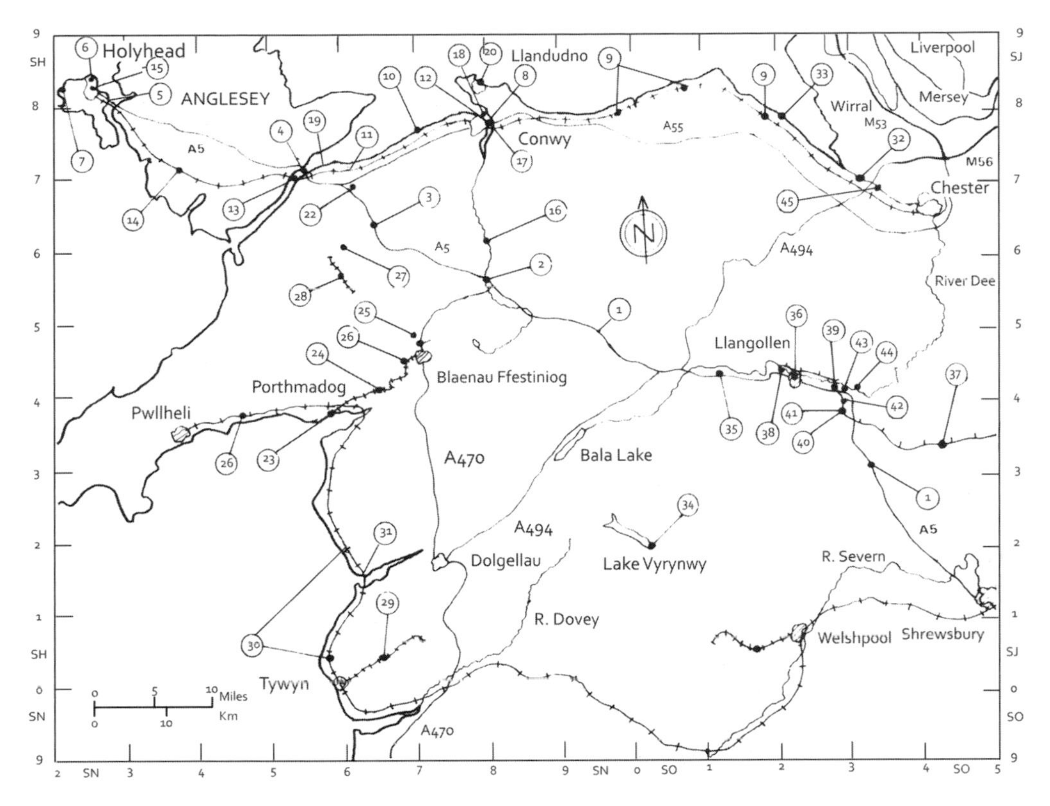

North Wales map.

N1. The Holyhead Road
N2. Waterloo Bridge, Betws-y-Coed
N3. Nant Ffrancon Pass
N4. Memai Bridge
N5. Stanley Embankment
N6. Holyhead Harbour
N7. South Stack Lighthouse
N8. Conwy Suspension Bridge
N9. Chester & Holyhead Railway
N10. Penmaenmawr Viaduct
N11. Ogwen Viaduct
N12. Conwy Tubular Bridge
N13. Britannia Bridge
N14. Malltraeth Viaduct
N15. Holyhead Railway Station
N16. Llanrwst Bridge
N17. Conwy Arch Bridge
N18. Conwy Tunnel
N19. Bangor Pier
N20. Llandudno Pier
N21. Colwyn Bay Pier
N22. Penrhyn Tramroad and Railway
N23. The Cob Porthmadog
N24. Ffestiniog Railway
N25. Blaenau Ffestiniog Railway Tunnel
N26. Ffestiniog pumped storage scheme
N27. Dinorwig pumped storage scheme
N28. Snowdon Mountain Railway
N29. Talyllyn Railway
N30. Cambrian Coast Railway
N31. Barmouth Viaduct
N32. Hawarden Swingbridge
N33. North Wales Coast Defences
N34. Vyrnwy Dam
N35. Upper Dee Bridges
N36. Llangollen Ancient Bridge
N37. Shropshire Canal, Llangollen
N38. Horseshoe Falls Weir
N39. Pontcysyllte Aqueduct
N40. Chirk Aqueduct
N41. Chirk Canal Tunnel
N42. Chirk Railway Viaduct
N43. Cefn Viaduct
N44. River Dee Viaduct
N45. River Dee Channel

The Dee Estuary is the northern end of the Welsh border with England skirting the west of the city of Chester. Westward lie the Vale of Llangollen, the Clwydian Range, the Vales of Clwyd and Conwy and the natural grandeur of Snowdonia, with the Lleyn Peninsula to the west, the Menai Strait and the Isle of Anglesey to the north and the Cambrian Mountains to the south.

North Wales is rich in mineral deposits, lead, coal and iron in the east, slate in Snowdonia, and non-ferrous metals, particularly copper, on Anglesey. These have given rise to industries associated with these raw materials and a communication network for both materials and finished products. The importance of the transport link to Ireland from Holyhead has required the construction of major east-west routes.

The mountainous landscape has presented serious obstacles to communications, resulting in many excellent examples of the science and art of civil engineering, in particular the works of two of the most eminent 19th-century engineers, Thomas Telford and Robert Stephenson. In constructing his highway through the mountains and on to Holyhead Telford built some of his finest works. Stephenson chose the coastal route for his railway from Chester to Holyhead; this was not without difficulties and his ingenuity was severely tested. John Rennie's work at Holyhead Harbour and Rendell's massive breakwater were essential to the development of the Irish connection.

92 *Three historic bridges at Conwy – the modern steel arch bridge, Telford's Suspension Bridge and Stephenson's Tubular Bridge. (© RCAHMW)*

North-West Wales

N1. The London to Holyhead Road, Welsh Section – Chirk to Holyhead (SJ 400179 to SH 250832). In 1810 Thomas Telford was commissioned to report on the state of the road from London to Holyhead and to suggest improvements. In 1815 Parliament voted funds for what was to become one of the finest achievements of one of the great British civil engineers.

In the last years of the 18th and the first years of the 19th centuries the road from London to Holyhead, for the sea crossing to Ireland, was second in importance only to that from London to Dover. But over much of its length it was in a dreadful state of repair, nowhere more so than across North Wales. On Anglesey, reached by ferry, much of the route was only a grass track, and through the Welsh mountains the road ran along the edge of unprotected precipices with gradients as steep as 1 in 6.5. Consequently travel was slow and dangerous. The Irish mail coach took almost 46 hours to travel from London to Holyhead averaging less than six miles per hour.

Parliament, through the Holyhead Road Commission, required that the road, then under the control of 23 separate turnpikes, should be improved over its 267-mile length. Telford was appointed to superintend the improvement works. Between London and Shrewsbury the works included the easing of gradients, widening, and short diversions in places.

West of Shrewsbury more drastic improvements were necessary and for this section Parliament authorised a Parliamentary Turnpike Commission to take over from the turnpike trusts. Telford's aim was that the road should not have gradients steeper than 1 in 30 to enable the coaches to attain a regular speed of about 10mph, although in a few places certain sections necessitated 1 in 22. This required substantial engineering of embankments and cuttings, much of which can be seen today.

By 1819 most of the 85 miles between Shrewsbury and Bangor had been made safe for traffic. Where the road passed through Glyn Diffwys, west of Corwen, the hillsides were blasted to enable it to be widened and regraded. This section has now been bypassed and retained as a footpath and cycleway. From Rhydllanfair on the River Conwy a three-mile stretch was built with a maximum gradient of 1 in 22. After crossing the Conwy on Waterloo Bridge into Betws-y-Coed the road followed the south bank of the River Llugwy to a point just upstream of Swallow Falls, where it crossed the river and took a new line for a mile before rejoining the old route to Capel Curig. A new alignment was chosen through the Nant Ffrancon Pass, through Bethesda and on to Bangor. Between 1820 and 1823 Telford built 20 miles of new road across Anglesey from Menai Bridge to Holyhead, including the Stanley Embankment between Anglesey and Holy Island. The turnpike system eventually lapsed in the 1880s although tolls continued to be collected on the Anglesey section until 1895.

Telford's route is now the A5 trunk road and in 1998 the Welsh Office designated it an Historic Route and erected special signs attributing its design to Thomas Telford, the intention being to limit any changes to it and to preserve its many original features. Relics in the shape of tollhouses, retaining walls, bridges and road furniture remain. A tollhouse, tollgate and milestone can also be seen at the Blists Hill Museum at Ironbridge (SJ 695032).

Telford's method of road construction included a pavement, seven inches deep at the centre and five inches at the sides, of large stones placed by hand as closely as possible, the upper layer being of stones no wider than three inches. The interstices were hand-packed with smaller stones. Then followed a further six-inch layer of stones of less than six ounces and less than 2½in size with a final surfacing of 1½in of gravel as a binding layer compacted by the traffic. Cross drains were provided at 100-yard intervals connecting with side ditches and the road had a cross camber of about four inches. Telford provided depots every 400 yards or so along the road to store material for repairs. Many of these can still be seen.

Telford's methods were more expensive than previous systems, including that being developed by John McAdam around the same time, but were more durable. Both Telford's and McAdam's principles of construction remain true to the present day. (HEW 1212)

N1.1. Tollhouses (SH 557713). Fifteen tollhouses were built between Holyhead and Shrewsbury to standard patterns. On the mainland they were four-roomed single-storey buildings but on Anglesey, for reasons at present unknown, they were two-storey. This fine example of Telford's two-storey toll houses at Llanfair P.G. still has its toll boards in place. It is octagonal with a slate roof and central chimney, as well as a slate-roofed canopy. There are three rooms downstairs and one upstairs. Others on Anglesey are at Caergeiliog (SH 315786), Gwalchmai (SH 398761) and Penrhos (SH 271804), the latter having been moved 100m from its original location. (HEW 0457)

Of the standard single-storey design used on the mainland several examples still exist, and the one at Ty Isaf is one of the best preserved (SJ 167422).

Opposite the tollhouse at Ty Isaf is an example of a weighbridge house. Weighing machines were provided at key locations on the route so that tolls could be charged according to the amount of wear caused to the road. The building housed the levers and weights for the platforms.

A very ornate example of the single-storey tollhouse can be seen at the east end of Conwy Suspension Bridge (SH785776), which Telford built in a medieval style to accord with the nearby castle.

N1.2. Tollgate (SH 557713). Recently carefully restored as part of refurbishment work to the Menai Bridge, this gate is one of at least 11 known still to exist. It is the most complete example of Telford's 'sunburst' design, retaining the original posts and latch, and stands alongside the bridge house at the south end of the bridge. Another is

93 *Llanfair P.G. tollhouse.*

94 *Ty Isaf tollhouse. (© NH)*

95 *Weighbridge House, Ty Isaf. (© BD)*

96 *(Left) Tollgate at Menai Bridge.*

97 *(Right) This milestone is outside the Thomas Telford Centre (Canolfan Thomas Telford), at Menai Bridge.*

on display in the Bridges Exhibition at Canolfan Thomas Telford, Menai Bridge, and one can be seen at the Conwy Suspension Bridge. (HEW 0458)

N1.3. Milestone (SH 276803). This milestone is one of the original stones on the Welsh section of the Holyhead Road and forms part of the Stanley Embankment Grade II listed structure. There were 84 such milestones between Holyhead and Chirk. A Cadw survey in 1998 of the Telford Road found all but five of these, although 39 had their cast-iron plates missing. The Highways Directorate of the National Assembly for Wales undertook a programme of refurbishment. Five new stones were sourced from a quarry with the closest geological match to the quarry at Red Wharf Bay, Anglesey, used by Telford and new plates were cast for all the missing ones. All were repainted to Telford's original specification. (HEW 0459)

N2. Waterloo Bridge, Betws-y-Coed (SH 798557) is one of Telford's major cast-iron bridges. This well-known bridge was built by him in 1815-16 to carry the Holyhead Road

98 *Waterloo Bridge. (© BD)*

over the Afon Conwy. Its single span of 105ft consists of five cast-iron arched girders at five-foot centres supporting cast-iron deck plates. The outer ribs carry the legend 'This arch was constructed in the same year the battle of Waterloo was fought', and the spandrels above are beautifully decorated with rose, thistle, shamrock and leek, modelled in relief by William Hazledine from his works at Plas Kynaston near Pontcysyllte. His foreman, William Stuttle, was responsible for the erection of the bridge. Telford standardised on spans of 105ft and 150ft although there is one exception of 170ft at Mythe Bridge, near Tewkesbury.

In 1923 the bridge was strengthened by concreting the three inner ribs and adding a seven-inch reinforced concrete deck. This was cantilevered to provide new footways and allow the road to be widened. In 1978 a new 10-inch reinforced concrete deck was added and the original cast-iron parapet fence, visible from outside, was protected by an additional fence inside. The masonry abutments were strengthened on both occasions. (HEW 0106)

N3. Nant Ffrancon Pass (SH 630660 to SH 730580) still has its original Telford masonry embankments and retaining walls. This part of the A5 Holyhead Road runs between Llandegai and Capel Curig and follows the Afon Ogwen east to the summit of Nant Ffrancon Pass, skirting the south side of Llyn Ogwen, and then follows the Afon Llugwy downstream to Capel Curig. During the 1990s seven phases of soil nailing were carried out to strengthen Telford's retaining walls in the Nant Ffrancon Valley while preserving the original appearance.

An older track constructed in 1791 by the Penrhyn Estate can still be seen on the west side of the Nant Ffrancon Valley. A later (1802) turnpike was built on the east side. This steep turnpike road has been overlain by the new road which was constructed with a maximum gradient of 1 in 22. (HEW 0460)

N4. Menai Bridge (SH 556715) is one of Telford's finest works and was the longest span suspension bridge in the world when built. In 1817 the Holyhead Road Commissioners instructed Telford to design a bridge to replace the ferry across the Menai Strait.

99 *View of Nant Ffrancon Pass from the south, looking towards Bangor. (© BD)*

100 *Pont Pen-y-Benglog, Nant Ffrancon Pass, over what is believed to be the old turnpike bridge. (© BD)*

The crossing by ferry was unpredictable and dangerous. Plans were ready by February 1818.

The construction took John Wilson, the contractor, a period of seven years from 1819. The bridge has an overall length of 1,000ft, seven stone approach spans and a main suspension span of 579ft carrying the road 100ft above sea level, an Admiralty requirement to allow passage of shipping. A full account of the design and construction is given by W.A. Provis, the resident engineer, in a large folio volume published in 1828.

The piers are faced with Penmon limestone (known as Anglesey marble) and the main towers are hollow between high-water and deck level; the other piers are solid. The deck was suspended from four sets of wrought-iron chains, the links for which were made by William Hazledine at Upton Magna Forge near Shrewsbury and tested in his Coleham workshops.

Following damage in a storm in January 1839 modifications to the bridge were made to increase the weight of the deck. In 1893 Sir Benjamin Baker replaced the timber deck with steel troughing on flat-bottomed rails. In 1940 the chains were replaced by two sets of steel chains, the deck rebuilt in steel to take heavier road traffic, and a cantilevered footway was added each side. Nevertheless, the modern alterations retain the gracefulness of Telford's original structure, set as it is in magnificent scenery. (HEW 0109)

N5. Stanley Embankment (SH 276803 to SH 285799) is a major feature of Telford's road. The old road crossed the waterway between the main island of Anglesey and Holy Island at the narrowest point, Four Mile Bridge, but Telford crossed it by the most

101 *Menai Bridge.*

direct route. The 1,300yd-long embankment is 16ft high and 114ft wide at its base with side slopes generally 1 in 3, which curve upwards into vertical parapet walls to protect the roadway. Between 1845 and 1848 Robert Stephenson superimposed his Chester & Holyhead Railway on a widened embankment on the south-west side, next to the road. Telford's work has been entirely encased by later repairs, including storm protection to the side slopes. The structure is listed Grade II.

The construction of the embankment with only a narrow bridge to allow tidal flow through it created the Inland Sea between the embankment and the original crossing further west at Four Mile Bridge. The changes to the tidal patterns each side or the embankment have created a variety of habitats ideal for a wide range of wildlife which led to the designation of the area as an SSSI in 1961. This designation meant that, when the new A55 embankment was built recently, additional compensatory habits and environmental works were necessary before the new road could be constructed. (HEW 1246)

N6. Holyhead Harbour (SH 250840) was an important link in the route to Ireland in the early 19th century. The 63-mile route from Holyhead to Dun Laoghaire is the most direct sea crossing from Great Britain to Dublin.

The history of the port is complicated because of the changing names of some of its elements. In the 17th and 18th centuries there was just a creek which dried out at low water, later known as the Inner Harbour. This was given protection by the Admiralty Pier completed in 1824, designed by John Rennie. It ran eastward from Salt Island and was linked to the town by a swing bridge over the sound. The Custom House and Harbour Office on Salt Island were part of the scheme. A triumphal arch designed by Thomas Harrison, a prominent architect and County Bridgemaster of Cheshire, commemorates the visit of George IV in 1821. The next stage comprised Telford's Dry Dock of 1825 opposite the tip of the Admiralty Pier and his South Pier in 1831. The Admiralty Pier then became known as the North Pier.

The dry dock was designed to drain at low-water spring tides but by 1829 steam pumping was installed to empty it over the neap tides. The Science Museum in London has a photograph of the Boulton & Watt engine that drove the two bucket pumps. Original drawings suggest that the dock was built with hinged gates but later a caisson was substituted. In 1939 a new caisson was fitted for wartime purposes. The dock is now derelict and has been filled, but the tops of the masonry walls are visible at the entrance.

The New Harbour was authorised in 1847, the area between the North and South Piers becoming the Old Harbour. A breakwater 7,860ft long, built with stone brought down by a railway from quarries on Holyhead Mountain, was under continuous construction from 1845 to 1873. The objective was to enclose a deep-water sheltered roadstead of 400 acres in addition to the 276 acres of harbour. The engineer was J.M. Rendel and on his death in 1856 the work was taken over by John Hawkshaw (later Sir John), assisted by Harrison Hayter who had already been engaged on the project for some years. The breakwater is still the longest in Britain.

The Chester & Holyhead Railway came in 1848, first with a terminus at the head of the creek, then in 1856 to the Admiralty Pier which, widened and extended in timber, became known as the Mail Pier as it was used by the Dublin Mail Boats from 1849 to 1925. The timber extension was demolished some time between 1935 and 1942.

Between 1875 and 1880 the London & North Western Railway Company developed the Inner Harbour with two quay walls forming a 'V' which accommodated the new passenger station. A new graving dock was built on the east side of the harbour. In 1922 the

102 *Holyhead Harbour and breakwaters in 1984. (The ferry terminal is now on the east side of the harbour, where the container cranes are in this aerial photograph). (© RCAHMW – Aerofilms Collection)*

channel to the station berth was deepened by a rock-breaking dredger. Previously Telford had used a diving bell to trim off submerged rock. A large part of the Inner Harbour is now redundant and separated from the main harbour by two low-level bridges.

In the past four decades there have been a number of developments at the port, including a deep-water berth for importing aluminium ore, and roll-on roll-off berths in the Inner Harbour. More recently the development of the high-speed catamaran ferries has led to the construction of a number of special berths and reclamation of significant areas of land. The Inner Harbour is now crossed by the striking pedestrian Celtic Gateway Bridge, giving direct access to the town centre along the causeway used by ferries to travel between the terminal collection area and the various berths. (HEW 1095)

N7. South Stack Lighthouse (SH 202823) is a 91ft-high circular masonry lighthouse. The north shore of Anglesey is roughly at the same latitude as Liverpool, above the general line of the North Wales coast. The Skerries Lighthouse offshore marks the main shipping lanes to the north-west. Holyhead Harbour is seven miles to the south and is marked on the west side by South Stack Lighthouse. There are in addition two local lights, one 48ft high on Admiralty Pier dating from 1821, the other 63ft high on the outer breakwater dating from 1873.

South Stack Lighthouse was designed by Daniel Alexander and first lit on 9 February 1809. Built from stone quarried on the site, it consists of a tower of traditional form, tapered and painted white, with its gallery and lantern about 90ft above the rock. The lighthouse is flanked by a long low building with a two-span pitched roof and three smaller buildings. These were the engine room and dwellings, but the operation is now automatic. When fog obscured the light a subsidiary light could be lowered down the face of South Stack to sea level.

Illumination was initially by Argand oil lamps, then paraffin vapour and finally electricity from 1938, when a separate fog station was built. The light is almost 200ft above mean high water and can be seen for over 28 miles. (HEW 0756/01)

N7.1. South Stack Bridge

The 90ft wide chasm between South Stack and Holy Island was first crossed by a hemp rope 70ft above sea level along which a sliding basket was drawn. On the Holy Island side this was approached by a flight of 400 steps. A donkey was used to carry

103 *South Stack lighthouse and bridge. (© BD)*

104 *The approach to Conwy bridge from the east. The tollhouse is on the right of the gate. (© NH)*

the loads across the gap. In 1828 this method was superseded by the construction of a five-foot-wide iron suspension bridge with masonry towers 115ft apart and a clear span of 108ft. This was demolished in 1964 and replaced with a bridge of aluminium alloy supported from the original towers. This bridge has been in turn replaced in 1997 by a lightweight truss bridge between the original towers. Part of the original chains can still be seen between the towers and the anchorages. (HEW 0756/02)

N8. Conwy Suspension Bridge (SH 785776) is the most successful of Telford's 'gothic' suspension bridges. Opened in 1826 as part of Telford's improvements to the Chester to Bangor road, this bridge has two pairs of solid ashlar limestone towers 12ft 4in in diameter and 40ft high supporting the chains. These span 327ft between the towers. The pairs of towers are linked by castellated walls containing the 10ft-wide carriageway arches to accord with the nearby medieval castle.

The wrought-iron chains are arranged in two tiers of five nine-foot-long links joined by deeper plates, the joints of which are alternated. Vertical rods at five-foot intervals carry the deck suspended from the junction plates. The present deck is a replacement dating from 1896. The original deck was probably made up of a light iron framework braced by bars on its underside, upon which were laid two longitudinal layers of fir planks. The bridge has been progressively strengthened, including in 1903 additional wire cables added above the chains. Later a footway was added on the north side to provide a six-foot-wide footpath, leaving the main bridge for vehicular traffic only.

Telford's bridge was in use until 1958 when the new bridge (**N17**) was built alongside. Plans to demolish it led to a public outcry and the bridge was taken into care by the National Trust. The footway was removed after the new road bridge was opened and Telford's bridge pedestrianised. A very ornate tollhouse and typical Telford 'sunburst' gate have been preserved at the eastern end. (HEW 0107)

N9. Chester & Holyhead Railway (SJ 413670 to SH 248822) is an early railway with many innovative features. The success of the Liverpool & Manchester Railway showed that rail travel offered speeds of over three times those possible on roads and effectively made Telford's improvements to the Holyhead Road out of date, but communications between Great Britain and Ireland remained important. Of the three railway schemes considered for this area the one adopted was that of the Chester & Holyhead Railway, incorporated in 1844 with Robert Stephenson, son of George, as engineer-in-chief and completed in 1850 over a route surveyed by his father in 1838.

Alexander MacKenzie Ross was Stephenson's assistant and among the main contractors were Edward Betts, William MacKenzie, Thomas Brassey, John Stephenson and Thomas Jackson. In 1847, with the opening of the line in prospect, Hedworth Lee, who had been Ross's assistant, was appointed resident engineer for the railway.

The line skirts the North Wales coast and for some 43 miles major heavy civil engineering works protected the railway and the coast from the sea. The new A55 road built in the 1990s faced similar problems and modern civil engineers have used solutions similar to those of Stephenson, although pre-cast concrete units were used to protect the embankment at Llandulais as an alternative to stone.

Important viaducts on the railway are the 22-arch Ogwen Viaduct at Talybont near Penrhyn Castle and at Malltraeth on Anglesey, but of prime importance are the river crossing at Conwy and the Britannia Bridge over the Menai Strait. Near its western terminus the railway crosses to Holy Island on a widening of Telford's Stanley Embankment.

Francis Thompson designed the original stations as two-storey rectangular structures in the Georgian style, one of the smallest of which is the first reached after crossing the Menai Strait and is the best-known to tourists, having one of the longest place-names (and station sign) in the world – Llanfairpwllgwynghyllgogerychwyrndrobwllllantysiliogogogoch (better known locally as Llanfair P.G).

The Chester & Holyhead Railway was built for speed, a feature emphasised in 1857 when John Ramsbottom installed the first ever water troughs near Mochdre, later re-sited to Aber (SH 652724). The troughs were several hundred yards long and located between the rails. By lowering a scoop the locomotives picked up water and this made non-stop long-distance rail travel a reality, avoiding the need to stop for rewatering. (HEW 1094)

N10. Penmaenmawr Viaduct (SH 696762) is a major viaduct on the Chester to Holyhead Railway. On 22 October 1846, during the construction of the sea wall at Penmaenmawr to protect the railway, a severe storm with 40ft waves destroyed the most exposed section of the wall in the presence of the Engineer, Robert Stephenson. He decided to replace it by an open viaduct some 182 yards long with 13 equal spans through which the waves could dissipate their energy up the sloping beach beneath.

The masonry piers, which were built in coffer-dams, are 32ft wide and 6ft 4in thick. They stand to a height of 41ft, about 15ft above high water, and were protected from scour by piles retaining large boulders set in concrete to form a pavement extending some 18ft seaward. The original deck consisted of four longitudinal cast-iron girders of inverted 'T'-section under each track, 14in wide, 4½in wide at the top, carrying timber way beams. The decking was of timber laid crosswise.

Work began in March 1847 and the viaduct opened to traffic on 1 May 1848. The contractors were Wharton & Warden. In 1908 the deck system was replaced by six-ring brick arches of 13ft rise. (HEW 0829)

105 *This view taken during construction of the new A55 highway shows the railway viaduct at sea level. Above this is the 1936 road replacing the Telford road, seen just above it. On the right is the start of construction work for the A55 tunnel through the headland. (© Welsh Assembly Government)*

N11. Ogwen Viaduct (SH 601707) is a 22-span viaduct by Robert Stephenson. Built as part of the Chester & Holyhead Railway in 1847, it was originally intended to be an embankment but at the request of Lord Penrhyn the proposal was changed to this 22-span stone and brick arch viaduct. It carries a double-track line over the River Ogwen. Each arch is 22ft 6in about 42ft above the river. The viaduct has stone abutments and piers. The construction contract formed part of contract no. 9, awarded to Thomas Jackson for the section from 1½ miles west of Aber to just short of Menai, let in May 1845. (HEW 1280)

N12. Conwy Tubular Bridge (SH 785774) was the first ever tubular bridge and a significant advance in bridge building. Built between 1846 and 1848 by Robert Stephenson to carry the Chester & Holyhead Railway over the River Conwy, it is very similar in design to his larger bridge on the same line over the Menai Strait, Britannia Bridge, which followed it and marked a significant advance in the art of bridge building. It was the result of much experimental work into the load-bearing properties of hollow beams.

Stephenson contemplated using suspension spans with deep rectangular trough-shaped stiffening girders but concluded that if the top of these were to be closed in the girder might be self-supporting. Studies by Professor Eaton Hodgkinson into the strength of materials and careful testing of large models by the structural engineer and shipbuilder William Fairbairn confirmed the feasibility of this approach.

The model testing was done for Britannia Bridge by Fairburn using, finally, a 1/6th-scale model 78ft long. Earlier smaller scale tests by Fairburn modelled circular and elliptical beams as well as the rectangular section finally chosen (on the basis that it would be easier to fabricate and had performed well in the tests). It was decided to test

106 *Conwy Tubular Bridge. (© RCAHMW)*

a model of Britannia Bridge because if that was successful then Conwy with its shorter span would not be a problem. The design approach was especially innovative in that the testing included the effects of wind and temperature variations. Timoshenko in his book on strengths of materials records this as a great advance in bridge design. The calculations by Hodgkinson were not complete at the time construction started but a few months later they confirmed the design.

Edwin Clark was the resident engineer. Contractors were Nowell, Hemingway & Pearson for the masonry and Garforth of Manchester and Ditchburn & Mare the ironwork. The bridge is a single span 400ft long formed by two parallel rectangular wrought-iron tubes, each weighing 1,300 tons. They were built ashore and then floated out on pontoons, to be raised in position onto stone abutments on each side of the river. Masonry towers were built on the abutments and topped with battlements and turrets to harmonise with the adjacent Conwy Castle.

Before the bridge was commissioned loading tests were carried out on both tubes using up to 300 tons of iron kentledge, which produced a central deflection of three inches. By way of comparison deflection under passage of an ordinary train is said to produce a deflection of only 1/8th inch. Two piers were added in 1899, reducing the span by 90ft to compensate for the increased weight of the trains then in service.

The same design was used for Britannia Bridge but since a disastrous fire there in the 1970s the Conwy bridge remains the only early example of this novel constructional form. (HEW 0108)

N13. Britannia Bridge, Menai Strait (SH 542710) was the second and longest tubular bridge at the time. Opened in 1850 to carry the Chester & Holyhead Railway over the Menai Strait, Robert Stephenson's Britannia Tubular Bridge and that at Conwy built two years earlier were significant advances in bridge design at the time. Stephenson, William Fairbairn and Professor Eaton Hodgkinson were the designers as described in **N12** above. Although Conwy was the first to be built the design had concentrated on the longer spans planned for Britannia Bridge on the basis that this could then prove the concept and be used at Conwy.

As at Conwy the tracks were carried within twin rectangular riveted tubes 14ft 8in wide by 23ft to 30ft overall depth, built up from wrought-iron plates as used in shipbuilding. The two main spans were each of 460ft flanked by side spans of 230ft. With spans greatly in excess of any previously constructed for a railway, the successful building of both these bridges was an outstanding achievement and advance in the use of wrought-iron structures. Prior to Conwy the maximum span had been less than 130ft.

The four tubes for the two main spans, each weighing 1,800 tons, were built on the Caernarfon shore and then floated out and jacked up 100ft onto the towers. The same procedure had been used at Conwy and was subsequently used by I.K. Brunel for his Chepstow and Saltash bridges. He stood beside his great friend and rival to watch the launching of the first Menai tube. The landward spans were constructed *in situ* on temporary staging.

All four spans for each track were connected end to end through the towers to form a 1,511ft-long continuous girder and so take advantage of the economy of material stemming from continuity. They were fixed at the central pier and on roller bearings at the side piers and abutments. Foundation work commenced in April 1846, the first tube floated out and positioned in November 1849 and a single line opened to traffic in March 1850, with full completion in October of that year. With its adherence to straight lines, its massive masonry abutments and three masonry towers rising above the bridge, the structure represents an outstanding example of symmetry, harmony and proportion. Resident engineer was Edwin Clark and contractors were Nowell, Hemingway & Pearson for the masonry and Garforth of Manchester and Charles Mare of Blackwall for the ironwork.

In May 1970 an accidental fire destroyed the protective tarred timber roof and the heat caused the tubes to tear apart into four separate spans with sag of up to 36in in the longer ones, rendering the bridge unusable. It was replaced by a clever practical and modern structure incorporating a road deck over the railway for a new section of the A5 trunk road, relieving the narrow Menai Bridge of heavy road traffic. Stephenson's original masonry piers were retained, complete with the monumental lions guarding the ends of the bridge.

The new main spans designed by Husband & Co. of Sheffield and built and erected by the Cleveland Bridge & Engineering Company, with steelwork supplied by Fairfield's of Chepstow, are steel arches with eight panels of 'N' truss spandrel bracing in each half arch (similar to the Victoria Bridge built some 70 years earlier over the Zambesi River). Each side span was divided into three spans built in reinforced concrete. (HEW 0110)

107 *The original Britainnia Bridge. (© SKJ collection)*

108 *The reconstructed bridge with the new road above the railway. (© RCAHMW)*

109 *A section of one of the original box girders has been preserved as a monument nearby on the mainland side of the bridge. (SH 545708) (© WD)*

110 *One of Stephenson's monumental lions on the bridge.*

111 *Malltraeth Viaduct. (© BD)*

N14. Malltraeth Viaduct (SH 414690) on Robert Stephenson's Chester & Holyhead Railway was built in 1847-8 and opened to traffic in August 1848. The 19-arch viaduct carries the double-track railway over an artificial embanked course of the River Cefni, two side drains, two roads and part of Malltraeth Marsh reclaimed from the sea under Acts of 1788, 1790, 1811 and 1859. Its arches are semi-circular brick with stone quoin faces and stone and concrete inverts in the water spans. The contractor was Edward Betts. The arches were pressure-grouted in 1961 and in 1969 the piers and arches were lined with reinforced concrete. In 1976 the spandrels were drilled to release water. The structure is listed Grade II. (HEW 1288)

112 *Holyhead Station.*

N15. Holyhead Railway Station (SH 248822) is the terminus of the Chester & Holyhead Railway. Holyhead Harbour Station opened in 1848 as a temporary station and was replaced by the permanent station nearer the harbour in 1851. Finally, in conjunction with the further development of the Inner Harbour by the London & North Western Railway, a large new station and hotel were built and opened by HRH the Prince of Wales in June 1880. The complex included an arrival platform, two departure platforms, a hotel, a motive power depot, carriage sheds and sidings, cattle lairage and rail loading accommodation, import and export sheds and a through-line to the Mail Pier. Little of this now remains except the truss roof over platforms 2 and 3. The layout was a 'V'-shape with the downside track and buildings along the west arm of the harbour and the upside along the east arm,

with a hotel at the south end of the 'V'. Of the original overall roof structures only the upside remains. It is a good example of the type of overall roof used by the LNWR at that time for several of their principal stations. It is Grade II listed.

The site is curved and tapering and the 64 roof trusses vary in span, the largest being 62ft. They are supported on a substantial screen wall on the one side and cast-iron columns with short wrought-iron lattice girders between them on the other. With a 1 in 2 roof slope and a camber of about a foot in the main tie, the trusses are a distinctive design with a clerestory carrying louvred ventilators. The main queen posts are braced by four tie rods and a ring. The station has now been connected at the north end to the new ferry terminal with an access link built across the Inner Harbour to the town centre. (HEW 1098)

N16. Llanrwst Bridge (SH 798615) is believed to be the first Palladian-styled bridge in the UK. Reputedly designed by Inigo Jones, who was a native of the district, this well-proportioned three-arch masonry bridge carries a road to Gwydir Castle over the Afon Conwy in a most attractive setting. Known as Pont Fawr, the Great Bridge or the Shaking Bridge (it vibrates if the parapet is struck at a point above the central arch), it was built in 1636 and has a total length of 169ft. The central arch has a span of 60ft and the side spans 45ft. It is 13ft 2in wide between the masonry parapets which carry on their outside faces coats of arms and the construction date. The bridge is a Scheduled Ancient Monument and listed Grade I.

Centrally placed on top of one parapet is a bronze sundial erected to mark the tercentenary of the bridge, unfortunately at a point where it is inadvisable to linger because of the traffic and the narrowness of the carriageway.

The construction of the segmental arch rings is unusual. Although the lower voussoir rings are of normal shape with a slight taper inwards, the 4in by 20in slabs forming the second ring are dressed and laid to the curve with their long concave face downwards on top of the voussoir ring. (HEW 0164)

113 *Llanrwst Bridge. (© NH)*

114 *Conway Arch Bridge from Conwy Quayside. (© BD)*

N17. Conwy Arch (SH 785775) is a 310ft-span modern steel arch bridge built in 1958 to relieve Telford's suspension bridge of the increasing weight and number of road vehicles. It is a reasonably successful solution to the problem of blending a modern structure into the historic background of the earlier Telford and Stephenson bridges and the World Heritage Castle and Town Walls.

The intention was eventually to widen the bridge to four lanes and the present bridge is only half of the project, which explains the difference between the north and south elevations. The upstream, south, side shows a simple spandrel-braced arch and an approach viaduct of short spans supported on columns. The castle and bridges are mostly viewed from downstream and from this direction the viaduct is masked by a long wing wall with a pilaster resembling an old-style cutwater and proportioned like the turret of a castle. From this springs a graceful arch which, combined with the flat vertical curve of the roadway, gives the impression of an ancient bridge. The simplicity of this bridge makes a satisfactory contrast with the towers of the other two bridges and the castle.

In the 1980s the widening proposal and a bypass close to the town walls were abandoned in favour of a more expensive tunnel further downriver to preserve the historic views. (HEW 0692)

N18. Conwy Tunnel (SH 776786 to SH 787782) was the first immersed tube tunnel in the UK. The tunnel forms part of the A55 high-speed dual carriageway between Chester and Holyhead and bypasses the town of Conwy with its narrow streets and medieval archway, which were major bottlenecks on this road. A tunnel option was chosen in preference to a less expensive bridge to preserve the views of the town, castle and historic bridges. The tunnel is sited downstream of the earlier crossings and includes a 260m-long cut and cover section on the eastern side, a 710m immersed tube tunnel of six massive prefabricated reinforced concrete sections under the river and a further 120m cut-and-cover section on the western side. Each concrete unit was 118m long, 24m wide and 10.5m high and weighed 30,000 tons. Each was the full tunnel cross-section of two carriageways and a central dividing concrete wall. All were cast

115 *The east portal of Conwy tunnel is in the foreground; the west portal is on the far left and the empty casting basin, later to become the marina, is in the far middle. (© Welsh Assembly Government)*

116 *Bangor Pier. (© BD)*

in a purpose-built casting basin on the west side of the river and floated out in turn, sunk into a prepared trench dredged in the riverbed and joined together to form the middle section of the tunnel which was then covered with protective rock armouring. The material excavated for the casting basin and the trench was used to reclaim land upstream of the town and develop a bird sanctuary. The cofferdam itself has been retained as a marina.

The £102 million contract was let to Costain-Tarmac Joint Venture in 1986; work started in November and the Crossing was opened by HM The Queen on 25 October 1991. (HEW 2069)

N19. Bangor Pier (SH 584732) is a late Victorian pier, said to be one of the three finest in Britain. Garth Pier at Bangor was designed by John James Webster and built in 1896 by Alfred Thorne of London. This 1,500ft promenading and embarkation pier was founded on cast-iron screw piles and its deck supported by steel lattice girders. These were replaced in the 1980s. The 24ft-wide deck is timber planking on timber joists. Ornate lighting columns are attached at intervals to the outside parapets and there are octagonal kiosks each side at widened bays along its length and at corners of the end platform. There is also a pavilion on the end platform. The pier was damaged in 1914 by a coaster that had broken away from its moorings. A temporary repair was carried out and a permanent one in 1921.The pier was considered to be unsafe in 1974 and was closed. Arfon Borough Council proposed its demolition but a campaign to retain it was successful. The pier was listed Grade II and ownership transferred to the City Council. Raising the funds for repairs took until 1982 and the work was completed in 1988, the pier being reopened by the Marquess of Anglesey on 7 May. (HEW 0427)

N20. Llandudno Pier (SH 784830). Llandudno was one of the few Victorian seaside towns built specifically as a holiday resort. Llandudno Pier was designed by Charles Henry Driver, (Sir) James Brunlees and Alexander McKerrow and built by John Dixon of Liverpool in 1876-7 and is considered one of the finest in Britain. The design is unusual in that it has a 45-degree turn about a third of the way along its length.

The 2,295ft-long structure is in two sections. The main pier is carried on a wrought-iron lattice girder framework supported on cast-iron columns and extends 1,234ft to a T-shaped pierhead 60ft wide. The deck is lined with four pairs of kiosks with three larger octagonal kiosks at the head. At the shore end an arm of the platform connected with an 1883-4 pavilion some distance from the shore and an extension past the *Grand Hotel*

117 *Llandudno Pier. (© BD)*

118 *Colwyn Bay Pier. (© NH)*

119 *Penrhyn Tramroad. (© BD)*

increased its length to 2,295ft. The pavilion had a projecting gable centre portion and recessed wings with apsidal ends and was fronted by a verandah. The pavilion, which housed in its basement what was at one time the largest indoor swimming pool in Britain, burnt down in February 1994. (HEW 0432)

N21. Colwyn Bay Pier (SH 853792) is a late Victorian pier with steel girders. Victoria Pier at Colwyn Bay was designed by Mayall & Littlewood of Manchester and was built by the Widnes Foundry Company in 1899-1900 using cast-iron screw piles. Another comparatively late pier, this used steel in its deck construction. A modern feature was the use of electric light for illumination. The original pavilion was set to the right of the deck with a walkway allowing access to the pierhead on the left. It was of a very high standard with typical ornate finishing. It had capacity to seat 2,500 people and held performances of many famous operatic and ballet groups. It had a balcony which extended around three of the four walls. A new 'Bijou' theatre was built in 1917 for light entertainment. Both were destroyed by fire in 1923, after which the pier was purchased by the local authority and was increased in length from 316ft to 750ft with a new pavilion. This also burnt down in 1933 but re-opened in 1934. The pier closed in the late 1950s, since when it has been sold to various companies, re-opened and closed several times and is still being refurbished with a view to another opening. (HEW 1285)

N22. Penrhyn Tramroad and Railway (SH 592730 to SH 616660) is an early 19th-century tramroad. The Cegin Viaduct on the Chester & Holyhead Railway crosses the line of the narrow-gauge Penrhyn Railway, which used to carry slate to Port Penrhyn from the quarries at Cefn-y-Parc near Bethesda some 6½ miles inland. The quarries were in use as early as 1580 and were served by packhorses. In 1782 large-scale working began and the road from Capel Curig was improved. In 1801 packhorses were superseded by a horse waggonway, designed by Benjamin Wyatt, one of the few in Wales to use edge rails and flanged wheels. The first rails were unusual, being of oval section. The tramroad had three inclined planes. The line was rebuilt to 1ft 10¾in gauge in 1874 and worked with steam locomotives, as many as 28 being at work at one time, mostly in the quarry itself, which is now served by road. It was designed by Charles Spooner and built by Richard Perry. From 1879 until 1958 the line carried workmen as passengers. It finally closed in 1965 and the track lifted. Much of the route is now a footpath and cycleway.

An excellent selection of relics from the railway is preserved in the Industrial Railway Museum, which opened in 1965 in the stable block of Penrhyn Castle (SH 603720). Some of the permanent way point and crossing items are particularly interesting. One of the narrow-gauge locomotives on display is the 0-4-0T Hunslet *Charles* of 1882 whose sisters *Linda* and *Blanche* of 1893 were transferred to the Ffestiniog Railway. Penrhyn Castle itself was designed by Thomas Hopper for G.H.D. Pennant and built between 1827 and 1837. It now belongs to the National Trust. (HEW 1257)

N23. The Cob, Porthmadog (SH 572384 to SH 584 378) was a major land reclamation project in the early 19th century. The construction of the embankment known locally as 'Y Cob' was sponsored by William Alexander Madocks (1773-1828), MP for Boston and a local landowner, to reclaim 7,000 acres of land from the tidal waters of Afon Glaslyn before the river enters Tremadog Bay.

The 1,400yd-long rockfill embankment, 90ft wide at its base, 18ft wide at the top and 21ft high, was built on rush matting between a hilly peninsula called Penrhyn-isaf at the eastern end and the island of Ynys Towyn at its western end where the harbour now is. It deflects the Afon Glaslyn, scouring out what became the harbour. It was

120 *A view of the cob before the latest widening. (© RCAHMW)*

constructed by tipping from two three foot gauge tracks carried on timber trestles. It was built by John Williams of Anglesey between 1808 and 1811.

Madocks had carried out a reclamation project in 1800 at his Penmorfa Estate to reclaim an inlet and over 1,000 acres of land. He employed an engineer, James Creasey, who had worked on the Bedford River. Here a two-mile-long embankment was built of sand with turf protection with a height varying from 11ft to 20ft. He then employed John Williams, who had worked on this scheme to develop a small model town of about 100 houses, Tremadog.

The Cob now carries the A487 trunk road between Porthmadog and Maentwrog and was widened in 2002 to provide a footway and cycleway to the north of the carriageway. At the same time the tolls were removed but the tollhouse can still be seen at the eastern end. The **Ffestiniog Railway (N24)** used the south side to connect to the harbour for slate exports when the railway was built in 1833-6. (HEW 1192)

N24. Ffestiniog Railway (SH 571384 to SH 701459) is a narrow-gauge railway constructed through difficult terrain. This 1ft 11½in gauge railway was designed by James Spooner and his two sons and was opened in 1836 to bring slate from the Blaenau Ffestiniog quarries to the harbour at Porthmadog. It descended 700ft in 13 miles, the maximum gradient being 1 in 80. Upgoing trains were drawn by horses for most of its length, except on inclined planes which were redesigned by Robert Stephenson with waterwheels to provide the power. Downhill the horses rode in a special car with the loaded wagons going down

121 *Tan-y-Bwlch cast-iron bridge. (© NH)*

122 *Blaenau Ffestiniog Station.*

under gravity. The line crosses the bay at Porthmadog on the Cob.

The carefully graded route necessitated embankments up to 60ft high, built of slate with almost vertical sides, and two tunnels at Garnedd and Moelwyn, the latter replacing an incline in 1844. Just west of Tan-y-Bwlch station the line crosses the B4410 on a neat little cast-iron arch bridge of 18ft skew and 12ft square span, built in 1854. (HEW 0257, SH 647415)

Steam traction was introduced in 1863 by James Spooner's son Charles and passenger traffic began in 1864. The Fairlie engine *Little Wonder* introduced in 1869 was the first narrow-gauge application of the Fairlie patent. The first bogie rolling stock in Britain was introduced on this line 1871. By the late 1890s the railway was carrying over 130,000 tons of slate annually.

The line was closed in 1946 but restoration work was begun by the Ffestiniog Railway Society, many of whose members rebuilt the railway in their own time, re-opening it from 1955 in stages to make it a highly successful tourist attraction.

The construction of the Ffestiniog power station (**N26**) in 1957 severed the line south of Tan-y-Grisiau and a diversion for that part of the line was needed. The diversion begins at SH 679422 just beyond Dduallt station, where it turns off to the east and makes a complete loop clockwise (the only railway spiral in Britain); it then passes over the original line just south of the station before turning north again to pass through a new tunnel at Moelwyn, opened in 1977, and thence along the west side of the Tan-y-Grisiau reservoir. The line was reopened throughout its length to a new terminus in Blaenau Ffestiniog in May 1982.

In 2008 the company completed another narrow-gauge line, the Welsh Highland Railway, from Caernarfon to Porthmadog, with a link from its terminus north of the town to the Ffestiniog Railway terminus on the south. (HEW 0647)

N25. Blaenau Ffestiniog Railway Tunnel (SH 688503 to SH 697469). This two mile 340 yard-long tunnel under the 1,712ft-high Moel Drynogydd near the Crimea Pass is the seventh longest railway tunnel in Britain and the longest single track one. It was designed initially by Hedworth Lee of the London & North Western Railway (LNWR) to take a narrow-gauge single-track line from Betws-y-Coed to Blaenau Ffestiniog to join up with the Ffestiniog Railway. Construction began on 6 December 1873 with the contractor Gethin Jones starting work on the northernmost of the three shafts used for the excavation of the tunnel. Eight headings were used from the bases of the shafts and the portals, and extremely hard rock was encountered, causing Jones to abandon the work. The tunnel was completed by direct labour under the supervision of William Smith, who succeeded Lee as district engineer at Bangor.

123 *The western portal of Blaenau Tunnel. (© NH)*

During construction, the LNWR decided to alter the line to standard gauge – although this meant dispensing with the connection to the Ffestiniog Railway and altering the work already done. The tunnel is 18ft 6in high and 16ft 6in wide and is straight except for short

124 *Gethin's Bridge. (© BD)*

curved sections at either end. Most of the tunnel is on an ascending or descending gradient of 1 in 660 except for a short level section at the summit near the southern end of the tunnel where the rail level is 673ft above that at Betws-y-Coed station. For most of its length the tunnel is unlined but there are eight short lined lengths, one being where the tunnel passes through slate waste. The linings are parallel walls of waste slate with brick soffits. Except for this section the tunnel was driven through very hard rock, alternating layers of slate and greenstone. The tunnel was opened to traffic on 22 July 1879. (HEW 0755)

On the same line is the Lledr Viaduct (SH 780539) near Betws-y-Coed, crossing the A470 road, Afon Lledr, an occupation road and woodland. It is popularly known as 'Gethin's Bridge' after Gethin Jones, the contractor who built it. (HEW 1303)

N26. Ffestiniog pumped storage scheme (SH 679445) was the first pumped storage hydroelectric power station in Britain. A pumped storage system holds water at a high level which is released to generate electricity on demand. Water is pumped back to

125 *Llyn Stwlan, the upper dam. (© First Hydro)*

the upper reservoir during off-peak periods and this proves an effective way of storing energy for reuse when demand rises.

The upper reservoir was formed by enlarging Llyn Stwlan with a concrete dam 1,250ft long and 110ft high; the lower reservoir, 1,033ft below, was created by damming the Afon Ystradau near the village of Tan-y-Grisiau. Here the concrete dam is 1,855ft long and 50ft high. From the upper reservoir two vertical shafts fall 640ft to two pressure tunnels which slope towards the power station for 3,750ft and then feed four steel pipes leading to the turbines and pumps a further 700ft away.

The four alternators, each of 90mW output when driven by the turbines, can operate as motors to drive the pumps. The latter are uncoupled when the station is generating. For pumping, the pumps are started by the turbines and when synchronisation of the frequency with the system has been reached, the alternators become driving motors and the water to the turbines is cut off.

The power station building on the west side of the Tan-y-Grisiau reservoir is of steel-framed construction faced in local stone. At 273ft long, 72ft wide and 66ft high, it is probably the largest stone building to be constructed in Wales since Harlech and Criccieth castles.

Great care was taken with landscaping. The spoil from Stwlan dam was placed in the reservoir and the face of Tan-y-Grisiau dam is concealed from view by rock from the excavations. General design was by the former Central Electricity Generating Board with Freeman Fox and Partners, in association with James Williamson and Partners, and Kennedy and Donkin, as consultants. The main contractors for the civil engineering work were the Cementation Co. Ltd and Sir Alfred McAlpine & Son Ltd. Work began early in 1957 and was completed in March 1963.

Several streams flow into the lower reservoir and their flow is released through regulating valves in the dam, monitored by weirs in the streams above and below the reservoir to ensure the reservoir level is kept such that water from the upper reservoir can be accommodated and enough retained for pumping upward. (HEW 1238)

N27. Dinorwig pumped storage scheme (SH 598607). On the opposite side of Llyn Peris from the lower terminus of the Snowdon Mountain Railway in Llanberis is the site of the largest pumped storage station in Europe and the third largest in the world. As noted in N26 above, pumped storage is in effect a means of storing energy. Electricity generated from base-load power stations at periods of low demand is used to pump water from the lower reservoir to the other at a higher level. At periods of peak demand, water is released from the upper reservoir, passing through turbines which drive electric generators feeding the national grid. This avoids the need to provide expensive thermal power stations solely to deal with peak loads, or for operating base-load stations at below their economic capacity in order that output can be increased to meet fluctuations in demand. Hydroelectric stations can be brought on power in seconds, which makes this system invaluable for dealing with sudden demands for electricity such as occur at the end of popular television programmes or on an unexpected breakdown on the grid.

The upper reservoir at Dinorwig is the existing lake of Marchlyn Mawr (HEW 1237, SH 617619), the capacity of which was raised by a 1,970ft-long rockfill dam 118ft high. This permits the water to rise 108ft above the original level to accommodate the operational fluctuation of 100ft. The upstream face of the dam is sealed by a layer of asphaltic concrete, only the second instance of such an application in Britain. In 2008 the dam was heightened by an extra 2.5m to increase the capacity of the upper lake and the generating capacity of the power station.

From Marchlyn Mawr, 5,575ft of 34½ft-diameter low-pressure tunnel leads on a slight gradient to a 33ft-diameter vertical shaft 1,475ft deep. From the foot of this shaft, a 33ft-diameter high-pressure tunnel leads to the power station about 2,200ft away, dividing into six smaller-diameter tunnels before reaching the six turbines.

The machinery is housed in nine man-made caverns under the Elidir Mountain, the largest, the main machinery hall, being 590ft long, 80ft wide and 197ft high. The nearby transformer hall is 530ft long, 80ft wide and 62ft high; there are other massive shafts and galleries for hydraulic and electrical control equipment. The six turbines can work in reverse as pumps with their alternators as motors. On average, pumping lasts six hours each night and generation five hours each day, with an average output of 1,680mW at 18000V from an installed capacity of 1,800mW. A load of up to 1,320mW can be picked up in 10 seconds. From the turbines, six 12ft-diameter tunnels lead in pairs to three 27ft-diameter tail race tunnels which discharge into Llyn Peris, some 2,000ft away and 1,630ft below Marchlyn Mawr.

The project was designed by the former Central Electricity Generating Board with James Williamson of Glasgow, in association with Binnie & Partners of London, as consultants. When the main underground works were let in November 1975, a record was set for the highest value civil engineering contract ever let in the United Kingdom. The contractors were a joint venture of Sir Alfred McAlpine & Son, Charles Brand & Son, and Conrad Zschokke for the underground works, and Gleeson Civil Engineering for Marchlyn Mawr Dam. The station was fully commissioned at the end of 1983 and formally opened by HRH the Prince of Wales on 9 May 1984.

Great care was taken to minimise the effect of the works on the environment, in the heart of Snowdonia. Almost all of the construction is underground; even the 400kV outgoing transmission cables are buried for a distance of six miles. The few administrative buildings above ground are of local stone, much of it from old quarry buildings. A good deal of the spoil from the excavations, which include a total of 10 miles of tunnels, together with heaps of slate waste

126, 127 *The main machine hall during excavation and also plant installation. (© First Hydro)*

that had disfigured the neighbourhood for years, were tipped into old quarries or the deeper parts of Llyn Peris. Visits to the underground power station are arranged from the 'Electric Mountain' visitor centre in Llanberis. (HEW 1236)

A short distance away in Llanberis is the 'de Winton' water wheel (SH 585602) which drove the machinery for the Dinorwig slate works. These buildings on the east side of the lake are now the National Slate Museum, part of the National Museum of Wales. With its 59ft 6in diameter, it is one of the largest water wheels in the United Kingdom. (HEW 1284)

N28. Snowdon Mountain Railway (SH 582597 to SH 609543) has the double distinction of being the only rack railway in Britain and having the highest station, at 3,493ft above sea level. It was designed by Douglas and Francis Fox, using the Swiss system, developed by Dr Roman Abt, and using Swiss-built locomotives. It was built by A.H. Holme and C.W. King of Liverpool between December 1894 and January 1896. Their men only worked a five-day week, unusual in those days, because of the arduous conditions on the mountainside. It was opened on 6 April 1896 but a derailment and fatal accident on the first day meant a need for modifications to the track and it was another year before it reopened. The line carried 12,000 passengers in 1897; it now carries more than 140,000 a year.

The line is a single track with passing loops at the three intermediate stations. It rises 3,140ft in a distance of four miles 1,100 yards from Llanberis to the summit station on Yr Wyddfa, the highest peak in Snowdonia. The average gradient is 1 in 7.8, with the steepest being 1 in 5.5 and the mildest 1 in 50. The line was designed to the European narrow gauge of 2ft 7½in for the running rails, which are attached to steel sleepers. Two central rack blades are bolted either side of steel chairs, which are in turn bolted centrally to each sleeper. These engage with a powered pinion on the engine to ensure traction on the steep gradients. Inverted 'L'-section guard rails were added either side of the rack blades following the derailment on the opening day.

There are 10 bridges, the largest being the Lower Viaduct over Afon Hwch with 12 arches spanning 30ft, one skew arch spanning 38ft 4in over a road. This structure is built on a gradient of 1 in 8.5. (HEW 1222)

128 *Railway showing rack and pinion system in centre of track. (© BD)*

N29. Talyllyn Railway (SH 586005 to SH 671064) is a narrow-gauge railway built to carry slate, now a tourist attraction. In the 1830s, slate began to be quarried at Bryn Eglwys, a remote site above the village of Abergynolwyn about 7½ miles north-east of Tywyn (Towyn) on the Welsh coast. Pack horses took the finished slates to Aberdyfi (Aberdovey) but, following the purchase of the quarries by William McConnel, in 1865 a railway line was planned from Bryn Eglwys to Aberdovey. After the opening of the Cambrian Railways (originally the Aberystwyth & Welch Coast Railway) between Aberdovey and Llwyngwril in 1863, the terminus of the line was changed to Tywyn to enable traffic to be interchanged with the new coastal railway. The line was opened in December 1866.

129 *Abergynolwyn Station.*

The Engineer of the line was James Swinton Spooner, whose father and brothers were associated with the construction and development of the Ffestiniog Railway. Built to a gauge of 2ft 3in (the minimum allowed by the Act), the line runs almost straight from Tywyn (Wharf) station, where there is a narrow gauge museum, to Abergynolwyn station, a distance of just over 6½ miles, climbing almost continuously along the south side of the valley of the Afon Fathew. Public passenger services terminated at Abergynolwyn but the line continued for a further three-quarters of a mile to the foot of the first of several inclines connecting with the quarry workings. The maximum gradient of the line is 1 in 60. There were several intermediate stations, the most important being at Tywyn (Pendre), where the workshops and engine sheds were situated and where the passenger services from Abergynolwyn terminated. Just west of Dolgoch station, the line crosses the Nant Dolgoch at SH 650 045 on a three span segmental arch brick viaduct about 50ft above the stream, the line's most impressive feature.

By 1911 the controlling interest in the railway had passed from the McConnel family to (Sir) Henry Haydn Jones, MP. The slate quarries finally closed in 1946 and after the death of the owner in 1950 the future of the line was in jeopardy, but a Preservation Society was formed, the first in Britain, and took over the running of the line, which is now a major tourist attraction. For a time L.T.C. Rolt, the author of several books on engineering history, acted as general manager. In 1976 an extension of the public passenger service beyond Abergynolwyn to a new terminus at Nant Gwernol was opened. Locomotive No.7 has been named *Tom Rolt* in recognition of his contribution to the rescue of the Talyllyn Railway. (HEW 1210)

N30. Cambrian Coast Railway (SH 375350 to SN 697980 and SN 585816) is a late railway with significant lengths of coastal protection. The Cambrian Coast line around Cardigan Bay connects Pwllheli, Criccieth, Porthmadog, Harlech, Barmouth, Tywyn and Aberystwyth to the rest of the railway network via Dyfi (Dovey) Junction and Machynlleth. It also forms a standard-gauge link between the now active narrow-gauge lines – the Great Little Trains of Wales – such as the Ffestiniog, Talyllyn and Vale of Rheidol.

Authorised in 1861-2 (Aberdovey to Porthmadog and Porthmadog to Pwllheli) and 1865 (Glandovey to Aberdovey) as the Aberystwyth & Welch Coast Railway, the Cambrian lines were opened in 1863-7. The 1862 Act also authorised construction of a line from Pwllheli to Port Dinllaen on the Lleyn Peninsula which was being considered as an alternative port to Holyhead but never progressed beyond consideration. Initially the main effort was directed from Machynlleth to Aberystwyth. The section from Aberdyfi to Tywyn opened in 1863 with a ferry crossing south over the Dyfi to Fairborne and Borth, and a branch line from Fairborne to Penmaenpool and Dolgellau opened in 1865 pending completion of the Barmouth link. The contractor, Thomas Savin, failed financially in 1866 and Henry Conybeare took over the works, completing all but the Barmouth viaduct. He was replaced at the end of 1866 by George Owen, engineer of the Cambrian Railways. The viaduct at Barmouth was completed and the line fully opened in October 1867.

The line extends from Dyfi (Dovey) Junction for 55 miles north to Pwllheli and about 15 miles south to Aberystwyth. From Pwllheli the railway skirts the sea briefly at Pwllheli, Afon Wen, Criccieth and Harlech; then from north of Barmouth to Tywyn it follows the coast almost continuously. Further stretches occur on both sides of the Dyfi estuary, making a total of about 30 miles which are subject to coastal hazards. (HEW 1227)

The most difficult sections to maintain are at Llanaber, where huge concrete blocks connected by chains form part of the railway protective works, and at Friog near Fairbourne (SH 607118), where the line is carried on a cliff shelf 100ft above the sea with the A493 road 100ft above it. Derailments due to landslides occurred in 1883 and again in 1933 when coaches fell onto the beach. It is now protected from rockfalls by an avalanche shelter built in 1936. This has a 12-inch reinforced concrete roof slab some 60 yards long supported on reinforced concrete arches. (HEW 1929)

N31. Barmouth Viaduct (SH 623151) is a rare survivor of a timber viaduct. Crossing the Mawddach estuary, south of Barmouth, it carries the Dyfi Junction to Pwllheli single-track line and a footway across the shifting sands of the estuary. These overlie about seven feet of gravel above thick peat with steeply shelving rock only at the northern end, a significant challenge for the Victorian civil engineers. A rock foundation was possible only at the northern end where there are two fixed spans of 37ft 9in. The navigable channel, which is near the north shore, was originally crossed by an unusual opening span which tilted and drew back over the track. This was replaced with a swing span of 136ft with a central pivot and a fixed span of 118ft, both hogback trusses carried on cylindrical piers, which replaced the original cast-iron screw piles. To the south of the channel there are 113 18ft-span openings of wrought-iron girders on timber pile trestles, which have either been replaced, or encased in glass fibre reinforced concrete sleeves, as a protection against attack by marine borer. (HEW 1167)

130 *Barmouth Viaduct from the north bank.*

North-East Wales

N32. Hawarden Swing Bridge (SJ 312694) was the largest swingbridge in the world when built. Three spans of steel hogback 'N' trusses carry two railway tracks over the River Dee. The two fixed spans are each 125ft and the swing section is a record length of 287ft, which gave a clear opening of 140ft. Movement was originally by hydraulic rams and chains to a 32ft-diameter circular girder. The 90ft landward end of the bridge ran on a quadrant rail. The bridge was fixed and the machinery removed in 1971. The

131 *Hawarden Bridge. (© NH)*

girders have a maximum depth of 32ft and are at 27ft 6in centres. There are cross girders at 17ft centres with two pairs of rail girder bearers.

The bridge was built in 1887-9 for the Chester to Connah's Quay section of the Manchester, Sheffield & Lincolnshire Railway (later renamed the Great Central Railway), which in spite of its name had a line as far into Wales as Wrexham. Francis Fox was the Engineer, Horseley Bridge Company and John Cochrane & Sons the contractors, and the bridge was opened on 3 August 1889 by the wife of W.E. Gladstone, after whose residence it was named. (HEW 0775)

N33 North Wales Coast Sea Defences (SJ 180790 to SH 660740) is a comprehensive series of protection works for the construction of the Chester & Holyhead Railway. About one-half of the entire length of the Chester & Holyhead Railway is beset with problems arising from river and tidal flooding, coastal erosion and storm damage. As a result numerous important engineering works have had to be built. Robert Stephenson was the initial designer for the route between 1846 and 1847, using the foreshore for the eastern section from Connah's Quay to Colwyn Bay as the most suitable for the line. Although requiring extensive sea defences and land reclamation it offered an easier construction than inland. The section west of Colwyn Bay presented different obstacles with rocky headlands down to sea level which required tunnelling and viaducts. Other engineers have been involved over the years to rebuild and redesign some of the works but the route is substantially as envisaged by Stephenson.

At Holywell 2½ miles of embankment have constantly to be replenished by tipping slag some 100 yards from the railway. There is another wall 3,900 yards long between Mostyn and Point of Ayr.

The Abergele railway embankment, known as The Cob, consists of 1,960 yards of shingle beach and groynes flanked by the 1,470-yard Rhuddlan Marsh wall, first embanked in 1800 by trustees and raised and improved when the London & North Western Railway took it over in 1880.

At Llysfaen, east of Penmaenrhos Tunnel, landslips have frequently occurred, until now only partly compensated for by the fact that they have produced material to replenish the beaches to the east. In the 1990s the problem here was transferred from rail to road. The A55 dual carriageway has been built to seaward of the railway and is protected by some 22,000 'dolos' units. These are specially shaped interlocking concrete blocks, weighing five tons each, placed on a 1 in 2 slope as facing to secondary armour of one ton of rockfill, with a smaller rock core and a filter membrane.

At Old Colwyn there is an extensive area of unstable ground below the railway. For 680 yards there are groynes, toe walls, larger walls, pitched slopes and wave breakers.

Further west, as far as Llanfairfechan, the problem is different. High cliffs forced the early tracks, Telford's later coach road, and then Stephenson's railway to make use of shelves, either on sea walls at the base or higher up on the cliffs themselves, and occasionally to burrow through headlands in tunnels. East of Penmaenmawr half a mile of railway is protected by a 20ft wall leading to Penmaenbach Tunnel. Part of this wall was destroyed in 1945 and was replaced to a new design in reinforced concrete.

The 253-yard Penmaenmawr Tunnel (SH 995805 to SJ 025778) is extended by avalanche shelters to guard against rockfalls from the cliffs towering above it. The Penmaenmawr Sea Viaduct is part of a 1,500-yard length of solid masonry sea wall up to 40ft high, located some 20ft above high-water level and surmounting a 1 in 30 pitched slope down to the beach (**N10**).

Coastal defence is an important branch of civil engineering, often dramatic and always expensive. However, failure of coastal defences can be even more expensive, as was demonstrated during February 1990 when a section of the defences failed at Towyn, flooding the town to a considerable depth. (HEW 1228)

N33.1. River Clwyd 13th-century navigation improvements (SH 995805 to SJ 025778). Although not a work of coastal defence, it is worth recording here that along this section of the Welsh coast at Rhyl took place what must be one of Britain's earliest river navigation works. In about 1277, as part of the work being carried out to refurbish Rhuddlan Castle for the King, the section of the River Clwyd between Rhuddlan and the sea was rendered navigable for seagoing ships. This was achieved by straightening the meandering course of the river over a length of 2¾ miles. To carry out this work, diggers were recruited from Lincolnshire, some 300 being brought to Rhuddlan under mounted guard. The work was carried out between 1277 and 1280, the wages of the diggers employed on the river works being recorded in the castle accounts.

N34. Vyrnwy Dam (SJ 018193) was the first gravity dam constructed in Britain. It was built to supply water to the City of Liverpool and marked the introduction to Britain of the high masonry dam. Authorised by the Liverpool Corporation Waterworks Act in 1880, work began in July 1881. Water from Lake Vyrnwy first reached Liverpool in 1891 and the works were formally opened by the Duke of Connaught in July 1892.

The ground conditions were ideal for the construction of the gravity dam, which is 1,172ft long and has a maximum height of 145ft from the foundation to the crest of the spillway section. At this point the dam is about 127ft wide across the base. This was the first high dam designed to act as a weir and so dispense with a separate spillway. It is built of Silurian slate on a buried rock bar at the end of a lake carved by a glacier and filled with moraine material. The rock bar was located at a point where the moraine covering was least in depth.

132 *Vyrnwy Dam.*

The engineers, George Deacon and Thomas Hawksley (the latter resigning some way into the project), adhered to the two fundamental principles of great weight and water-tightness, and much care was taken to achieve these objectives. The mass of the dam consists of large irregularly shaped stone blocks, up to 10 tons in weight, set close together and bedded on cement mortar. The spaces between them were then filled with mortar into which smaller broken stones were forced. An important feature of the design was the provision of a drainage system in the foundations to prevent any build-up of water pressure on the underside of the base, possibly leading to overturning. (HEW 0214)

N34.1. Vyrnwy Aqueduct (SJ 0132202 to SJ 470938) is 68 miles long and there are now three 42in-diameter pipes to deliver up to 50 million gallons per day to Prescot service reservoirs east of Liverpool. The route follows the Dee-Severn watershed to maintain high ground until the basins of the Mersey and Weaver are reached. The first two pipelines were generally of cast iron but the use of riveted steel pipes to facilitate maintenance in the nine-foot-diameter cast-iron tunnel under the Mersey marked an early use of steel for trunk water mains.

The first three tunnels, at Hirnant, Cynynion and Llanforda, are alike, with brick and concrete linings to protect against leakage. Hirnant was later duplicated at Aber to enable maintenance to be carried out.

At the Oswestry reservoir, where the water is filtered, a 500-yard-long earth dam impounds some 52 million gallons storage. Two booster pumps at Oswestry were refurbished in 1991. There are several balancing reservoirs and water tanks. The largest tank, at Malpas, has a capacity of 4.5 million gallons, while the Norton tank of 650,000 gallons capacity is housed within a monumental sandstone tower some 100ft high.

The first section of the third pipe was laid in 1926-38 in steel. This saw the beginning of the more general use of bituminous-coated steel pipes for trunk water mains in place of cast iron, which had been in use since 1810. After 1946, to increase capacity, a fourth pipeline was laid upstream of Oswestry, with three booster stations downstream. The pipe crossings under the Mersey and the Manchester Ship Canal were rearranged in 1978-81. (HEW 1147)

N35. Upper Dee Bridges

There are several interesting bridges over the River Dee in this area.

N35.1. Pont Carrog (SJ 115437) is a substantial mid-17th-century five-arch masonry bridge on the upper Dee. There is a date on the eastern parapet of 1661. It is constructed of local stone and carries the B5436 Llidiart y Parc to Carrog road over the River Dee. It has five segmental arches and the cutwaters extend up to the parapet walls to form recesses at road level. Its total length is 186ft and the roadway extends the full width of 12ft between the parapet walls. The bridge is listed Grade II*. (HEW 0682)

133 *Pont Carrog. (© BD)*

N35.2. Corwen Bridge (SJ 069434) is the longest on the upper Dee. Built in 1704, it has six spans over the river and a total length of 107 yards. The date is inscribed on the downstream parapet but this may refer to the date of widening or original construction as it is recorded that a bridge existed on this site as early as 1577. The bridge is listed Grade II*. (HEW 1646)

N35.3. Pont Dyfrydwy (SJ 052412) near Cynwyd is an imposing bridge with four segmental masonry arches dating from about 1700. It carries a minor road between Cynwyd and Maerdy. This

bridge is a scheduled ancient monument and listed Grade II. (HEW 1645)

N35.4. Pont Cilan (SJ 021 374), about three miles upriver from Pont Dyfrydwy, carries a minor road from Llandrillo and was constructed at about the same period. It has two 40ft-span masonry arches and is listed Grade II. (HEW 1644)

N36. Llangollen Ancient Bridge (SJ 215422) is an example of early road engineering and is sometimes described as one of the Jewels of Wales. A bridge, built on this site in 1282, was reconstructed in sandstone about 1500 to the present style with four arches, three of which are pointed and the fourth segmental. It has piers with cutwaters that extend up to the parapet walls, forming recesses. In 1865 it was extended across the new railway by an additional span of iron girders and dressed freestone.

134 *Llangollen Bridge. (© BD)*

The bridge was widened on the upstream side from its original 12ft in 1873 to 20ft between the parapets and again in 1969 to 36ft. It is a scheduled ancient monument and listed Grade I. (HEW 0165)

N37. Shropshire Union Canal, Llangollen Branch (SJ 196433 to SJ 370318) is an important early canal with several significant civil engineering achievements, now a World Heritage Site. The Ellesmere Canal was authorised in 1793. Thomas Telford, then aged 36 and County Surveyor of Shropshire, was, to use his own words, appointed Sole Agent, Architect and Engineer to the Canal, with William Jessop as consulting engineer. It was proposed that the canal would link the Mersey, Dee and Severn Rivers via Ellesmere Port and Chester, passing through the North Wales coalfield and by way of Wrexham, Ruabon, Chirk and Frankton to Shrewsbury.

Work began on the 8¾-mile stretch from Chester to Ellesmere Port, and the Wirral Line opened in 1797. Work began at the same time on what was to be part of the main line but eventually became the Llangollen Branch. In 1800 a less expensive route was adopted from the Chester Canal at Hurleston (SJ 626553) to Frankton (SJ 370318), and the section north of Ruabon was never built. The Ellesmere Canal never reached Shrewsbury, the canal petering out south of Frankton. The Ellesmere company merged with the Chester Canal in 1813 to become the Chester & Ellesmere Canal.

By the time the decision had been made to change the originally projected north-south route of the Ellesmere Canal to a more easterly route, progress had been made up the Dee valley towards Llangollen, including the building of the Pontcysyllte Aqueduct to take the canal across the river at a high level. This section of canal, from Frankton to Trevor, therefore became a long branch and the originally proposed branch, from Frankton to the Chester Canal at Hurleston via Whitchurch, became the main line.

The branch was completed in 1805 under an Act of 1804 with an extension to Llantysilio, just to the west of Llangollen. This was done to obtain an adequate water supply from the Dee at Horseshoe Falls, which is fed along the main line of the canal to the reservoir at Hurleston (SJ 626553) at the junction with the Chester Canal, a fall of some 123ft. The Chester Canal had drawn its supply from a feeder above Bunbury Top Lock, but this was scarcely sufficient.

It is interesting to speculate that if the change of route had been decided upon earlier there might have been no need to build the magnificent Pontcysyllte Aqueduct, as the water supply could have been taken down the right bank of the Dee instead of the left. (HEW 1204)

N38 Horseshoe Falls Weir, Llantysilio (SJ 196433) is 460ft-long masonry weir, part of one of the earliest river regulation schemes in Britain. It was built by Thomas Telford between 1804 and 1808 to secure a water supply from the River Dee at Llantysilio to feed the Ellesmere Canal. It is not clear whether the name derives from the shape of the weir, which is actually J-shaped, from the horseshoe bend in the river, or even from the correspondingly named bend to the south above Llangollen on Telford's Holyhead Road. From this road there is an excellent view of the weir and its picturesque surroundings.

The weir is sited above rapids which are a venue for canoeing events. It is of masonry, 460ft long with an upstream slope and a vertical downstream face four feet high. The crest is of four-foot-square stones with a flat bull-nosed cast-iron capping in nine-foot lengths, each length secured to the masonry by three lugs.

135 *Trevor Basin, Llangollen Canal.*

136 *Horseshoe Weir.*

137 *Pontcysyllte Aqueduct. (© RCAHMW)*

In addition to the construction of the Horseshoe Weir, in 1808 Telford obtained permission from the local landowner to raise the level of Llyn Tegid (Bala Lake) many miles upriver by 2ft 6in using a weir and sluices to ensure an adequate reservoir. Similar arrangements were made at Llyn Arenig Fach and Llyn Arenig Fawr.

Much of the water from the River Dee now goes to water treatment plants for public supplies as well as for canal impounding. In more recent years the security of supply for Deeside and Merseyside has been augmented by the construction of reservoirs at Llyn Celyn (SH 936 356) in 1965 and Llyn Brenig (SH 822 426) in 1976. (HEW 1235)

N39 Pontcysyllte Aqueduct (SJ 271420) is probably the most spectacular example of canal engineering in Britain, and part of the World Heritage Site. It carries the Llangollen Branch of the Ellesmere Canal over the River Dee two miles west of Ruabon and was commenced in 1795 and completed in 1805. It was scheduled as an ancient monument in 1958, and is listed Grade I.

By far the greatest problem was the crossing of the River Dee and the original plans proposed a very expensive flight of locks down and up each side of the valley, which also would create water supply problems. Telford's innovative proposal was accepted as an economical alternative although nothing of this magnitude had been constructed before.

The project attracted great admiration throughout the whole country. Robert Southey wrote of 'Telford who o'er the Vale of Cambrian Dee aloft in Air at giddy height upborne carried his navigable road', while Sir Walter Scott described it as the greatest work of art he had ever seen.

138 *Detail of Pontcysyllte aqueduct.*

Before Telford's pioneering cast-iron aqueduct was built on the Shrewsbury Canal at Longdon-on-Tern, canal aqueducts had generally consisted of a heavy puddled clay waterway constructed on squat brick arches. Telford's scheme for the crossing of the River Dee developed the use of cast iron. At Pontcysyllte, a trough 11ft 9¾in wide is carried on four arch ribs over each of the 19 44ft 6in spans between the masonry piers. The aqueduct is 1007ft long and at the highest point is 127ft above the river. The towpath, which has a protective iron parapet railing, overhangs the channel on the east side, giving space for the movement of water displaced by the passage of boats while leaving a clear width of 7ft 10in. This is in contrast with Telford's earlier iron aqueduct at Longdon-on-Tern where the towpath is carried alongside the trough, level with the trough base. The slender masonry piers, which are partly hollow, taper upwards to 13ft by 7ft 6in at the top. Work started in 1795 and the aqueduct was opened in November 1805. There is a commemorative tablet fixed to the base of the pier adjoining the south bank of the river. The embankment at the south end should not be overlooked as it was one of the greatest earthworks undertaken at the time.

The aqueduct is important historically in that it brought together Thomas Telford, the designer of the aqueduct, William Jessop, the Engineer to the Ellesmere Canal Company (who approved the design), Mathew Davidson, who was Telford's supervising engineer, William Hazledine, the local ironmaster who supplied the ironwork from his new works at Plas Kynaston half a mile away, and the two master masons John Simpson and John Wilson. Members of this team were subsequently engaged on many famous civil engineering works, ranging from the Caledonian Canal to the Menai Bridge. (HEW 0112)

139 *Pontcysyllte Ancient Bridge and Aqueduct. (© RCAHMW)*

N39.1 Many pictures of the aqueduct show in the foreground the **Cysyllte Ancient Bridge** (SJ 268420), known locally as Pont Cysylltau, a three-span sandstone arched bridge across a bend in the river, built in 1697.

The superstructure consists of three segmental arches with double arch-rings built in two orders. It has slightly projecting keystones that make it one of the earliest examples of this method of construction popular in the 19th century. The main structure is squared block in-course buff sandstone but the approach walls are of uncoursed random rubble buff sandstone. (HEW 0160)

N40. Chirk Aqueduct (SJ 286371) is one of Telford's important aqueducts. This major work carries the Llangollen Branch of the Ellesmere Canal some 70ft above the valley of the River Ceiriog. This aqueduct has always been overshadowed by Pontcysyllte but in reality it is itself a major canal work.

Given that the Pontcysyllte aqueduct would take a long time to build, the company were anxious to open a line to Chirk and the iron and lime works there from a junction at Frankton. At first Telford had John Dunscombe as his assistant but in 1794 he secured Mathew Davidson as his Inspector of Works. The first plans were drawn up in 1794, the foundation laid in 1796 and the aqueduct opened in 1801. William Hazledine supplied the ironwork.

The structure appears from the exterior to consist of masonry piers and arches giving clear spans of 40ft and is 710ft long overall. However, Telford used cast-iron plates to form the bed of the five-foot-deep channel, thus enabling him to reduce the depth of the masonry, and hence the weight, above the piers. The plated bed was flanged at the edges and secured by

140 *The aqueduct with the later railway viaduct on the left. (© BD)*

141 *Chirk railway viaduct and canal.*

nuts and bolts at each joint and in addition was built into the masonry at each side. Side plates were added *c.*1870.

The sides of the waterway were waterproofed by ashlar masonry and hard burnt bricks in Parker's cement, obviating the need for the clay puddle used by the earlier canal engineers. (HEW 0111)

N41. Chirk Canal Tunnel (SJ 285374 to SJ 284378) is the longest of three tunnels on the Llangollen Branch at 459 yards long.It was built at the north end of Chirk Aqueduct by Thomas Telford in 1801. All three tunnels on the Llangollen Branch were unusual for the time in having a towpath taken through the bore, so boats no longer had to be 'legged' through the tunnel. The waterway is 9ft 6in wide with a five-foot towpath. The height above water level is 10ft. The portals make the tunnel appear larger than it is, as the bore is flared at the ends. It is believed that the ends of this brick tunnel were built in cut and the main length excavated from vertical shafts. The tunnel is listed Grade II*. (HEW 0162)

N42. Chirk Railway Viaduct (SJ 286372) was built adjacent to the canal aqueduct in 1848 for the Shrewsbury & Chester Railway to the designs of Henry Robertson by Brassey, Mackenzie and Stephenson with George Meakin as Agent. Originally with approach arches in laminated timber which were rebuilt in 1859 in brick as the rest of the viaduct, it consists of 16 segmental brick arches with abutments, piers, parapets and facings of stone to carry the double line of the railway over the River Ceiriog at the Welsh/English border. (HEW 0602)

N43. Cefn Viaduct (SJ 285412) was claimed to be the longest viaduct in Britain when built in 1848. Known also as the Dee or Newbridge viaduct this handsome structure, built by Brassey, Mackenzie and Stephenson to the designs of Henry Robertson in 1846-8, carries the double line of the former Shrewsbury & Chester Railway (later the Great Western Railway, although it never became broad gauge) across the River Dee and adjacent fields about a mile downstream from the Pontcysyllte Aqueduct. The

142 *Cefn Viaduct. (© RC)*

abutments and piers are of stone and the arches are of brick with stone facings. There are 19 openings of 60ft span and two of 15ft. The total length is 510 yards and the greatest height 148ft. (HEW 0568)

143 *River Dee Viaduct. (© Welsh Assembly Government)*

N44. River Dee Viaduct (SJ 300410) was claimed to be the highest road-over-river bridge in Britain. This concrete box-girder structure was erected in 1987-90 for the Welsh Office to carry the Newbridge bypass (A483 trunk road) over the valley of the River Dee not far from Chirk. The viaduct is 348m long and has five spans, the central span being 83m and the deck is 12.3m wide. The four reinforced concrete piers are all hollow and up to 55m in height; they support a post-tensioned prestressed concrete deck constructed on the balanced cantilever principle using variable depth single-cell box girders of web depth three to 5.7 metres and six metres wide. The piers are built on reinforced concrete foundations four metres thick on mass concrete six metres thick founded on bedrock. The viaduct was designed by Travers Morgan & Partners and Robert Benaim & Associates, who submitted an alternative deck design for the contractor Edmund Nuttall Ltd. (HEW 1782)

N45. River Dee Channel (SJ 397667 to SJ 275720) is an early 18th-century river improvement. The earliest improvement works to the Dee Estuary date from Roman times to improve access by sea to Chester when the river was improved *c.* A.D. 50 up to Farndon (SJ 412544). The port traded in stone, pottery, slate, tiles and timber.

Chester was the main north-west port until at least 1400 but siltation was reducing its trade, first being mentioned as a problem in 1445. The main shipping activities then moved to newly created wharves downstream at New Haven (Little Neston) and later Parkgate (SJ 280778). In 1558 Chester was named as a 'head' port for the region, with ports also existing on the estuary at Connah's Quay, Mostyn, Point of Ayr and Rhyl.

A new cut was authorised by Act of Parliament in 1699 to make the river navigable for 100-ton vessels, financed by tolls, but did not proceed. Work started in 1707 to improve the existing channel for vessels of up to eight-foot draft. A further Act in 1732 gave authority to Manley of Leeds with Nathaniel Kinderley as engineer to construct a new channel for 200-ton vessels, 80ft wide and eight feet deep. A channel eight miles long was constructed in four years from 1734. Excavation was by plough and eroders, with surplus water pumped out by portable windmills. The face of the bank was pitched with stone. About 1 million cubic yards of material were excavated. The new channel obstructed existing fords, and ferries were established at Saltney and Queensferry. In 1740 the River Dee Company was set up. The marshes silted up and were enclosed in stages from about 1740. Reclamation work continued into the 20th century.

The estuary continued to silt and several eminent engineers advised the Company, including Telford. In 1837 Sir John Rennie proposed a ship canal for 600/700-ton vessels but the work was not carried out. Nor was a proposal by Robert Stevenson and his son in 1838. In 1866 the main sea embankment was extended from Connah's Quay by 1,200 yards to Burton Point. The Company was wound up in 1889 and replaced by the River Dee Conservancy Board. The port continued to silt up and was eventually closed in the mid-20th century. (HEW 2047)

The area covered by this section extends eastwards from Cardigan Bay to the English Border and is bounded by the River Dyfi (Dovey) to the north and the mountain ranges of Brecknockshire and Ceredigion to the south. Predominantly a land of mountains and rivers, it is the source in the Plynlimon Ranges of the Severn and the Wye Rivers which flow eastwards through the region.

The landscape and high rainfall have been used to a great extent by water supply engineers, and man-made reservoirs and lakes abound. The Elan Valley in particular supplies water to Birmingham, and dams on the Tywi and Clywedog impound water as a means of regulating river flows for water abstraction at towns further downstream. In the west, non-ferrous metal mining has played a large part in this predominantly agricultural area, ores being exported from the coastal harbours. Limestone and agricultural products, cattle, sheep and wool have been transported eastwards. Most of the civil engineering in this area is concerned therefore with transportation, both rail and road, and water supply.

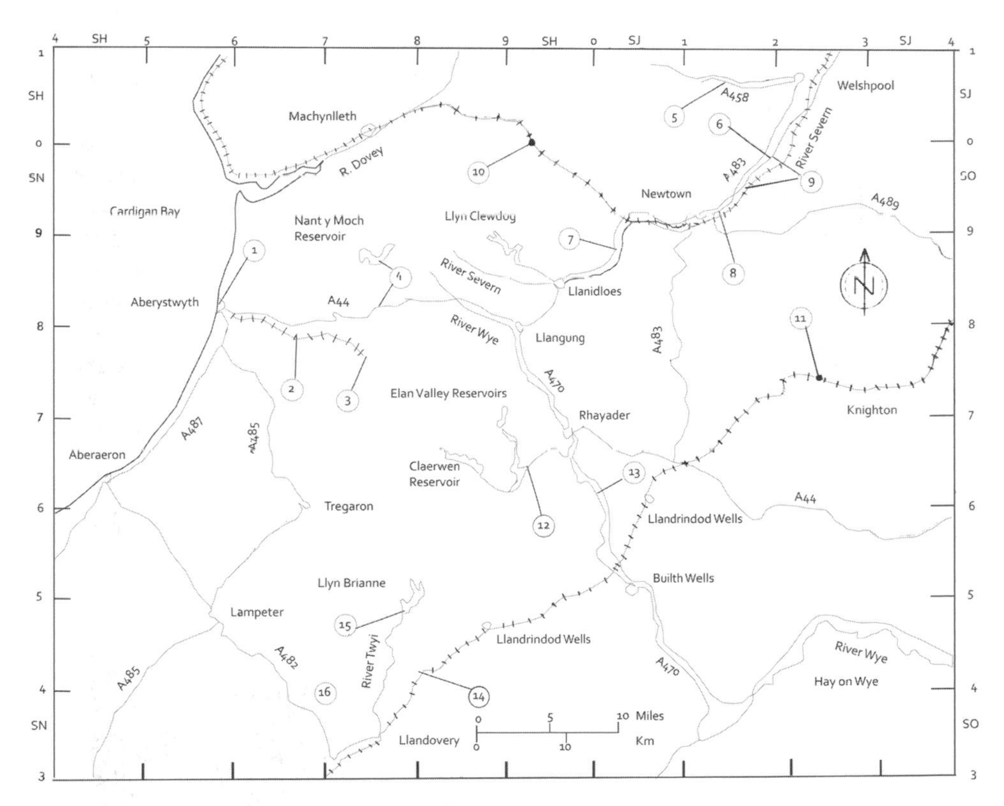

Map of Mid Wales.

- M1. Aberystwyth Cliff Railway
- M2. Vale of Rheidol Railway
- M3. Devil's Bridge
- M4. Cwm Rheidol hydro scheme
- M5. Welshpool & Llanfair Railway
- M6. Montgomeryshire Canal
- M7. Llandinam Bridge
- M8. Abermule Bridge
- M9. Caerhowell Bridge
- M10. Talerddig Cutting
- M11. Knucklas (Cnwclas) Viaduct
- M12. Elan Valley Reservoirs
- M13. Doldowlod (Ystrad) Bridge
- M14. Cynghordy Viaduct
- M15. Llyn Brianne Dam
- M16 Dolauhirion Bridge

M1. Aberystwyth Cliff Railway (SN 583826) is a cable-drawn balanced passenger railway, the longest in Britain. It was built at the northern end of Aberystwyth promenade for the Aberystwyth Improvement Company and opened on 1 August 1896. It is a funicular railway. Two passenger cars run on parallel 4ft 8½in gauge tracks and each is attached to a continuous wire rope cable that passes several times round a drum mounted on a vertical axis and operated by a stationary electric motor. The cars have a capacity of about 30 people. Their weights are mutually balancing, one car being raised as the other is lowered. Until 1921 the railway was operated by a water balance system, water being pumped to the upper station by a steam engine in the lower station.

Substantial earthworks were involved in achieving an approximately uniform gradient for the track, which rises nearly 400ft over its length of 778ft, a gradient of about 1 in 2 except at the ends where it changes to assist acceleration and deceleration. At the same time extensive development of gardens and pleasure grounds were provided. The railway was designed by G. Croydon Marks, later Baron Marks of Woolwich. (HEW 1130)

M2. Vale of Rheidol Railway (SN 585816 to SN 739770), 11¾ miles long, is a 1ft 11½in gauge line built to transport lead and zinc ore from the area around Devil's Bridge to the harbour at Aberystwyth. It starts about 680ft above sea level. The line now terminates at Aberystwyth Station, the harbour extension being removed when goods traffic ceased. The engineer for the line was (Sir) James Szlumper, the contractors were Messrs Pethick Brothers. It opened to goods traffic in August 1902 and to passengers in December.

Ownership of the line passed to the Cambrian Railways Company in 1913 and to the Great Western Railway in 1922. The GWR

144 and 145
View of line and locomotive. (Owen Gibbs Collection)

146 *Devil's Bridge. (© Owen Gibbs Collection)*

built two new locomotives and reconstructed one received from the former owners. Freight traffic ceased soon after and in 1931 the winter passenger service was withdrawn. It became part of British Railways Western Region in 1948 and in the 1960s as steam locomotives were phased out elsewhere it became British Rail's only steam-operated line. It was transferred to private ownership in 1989.

It is now solely a passenger line operating during spring and summer when it carries over 180,000 tourists a year through the scenic Rheidol Valley. Haulage is normally by steam locomotives and the three are named *Prince of Wales*, *Owain Glyndwr* and *Llewelyn*. There is also a diesel hydraulic locomotive. (HEW 1121)

M3. Devil's Bridge (SN 742771). Situated near Devil's Bridge Station on the Vale of Rheidol Line, Devil's Bridge is a popular tourist attraction. Three bridges cross the Afon Mynach, one above the other. The first, lowest, is a medieval masonry pointed arch of 15ft span said to have been built by the monks of Strata Florida Abbey, possibly in the 12th century but more likely in the 14th. The second bridge is considered to date from the mid-18th century and is built about 12ft above the original bridge with a flat segmental arch of 32ft span. The abutments and wing walls were taken down to rock by excavating behind those of the earlier bridge.

Early in the 19th century the spandrel walls surrounding this arch were raised by about six feet to allow the approaches to be less steep. At the beginning of the 20th century a steel bridge was built some seven feet above the earlier structures. This bridge, repaired and strengthened in 1971, continues to carry road traffic. The bridges are listed Grade II.

Access into the deep gorge is through turnstiles (for which a charge is made), via steps and zigzag pathways from which the bridges and the nearby waterfalls can be seen. (HEW 1120)

M4. Cwm Rheidol hydroelectric scheme (SN 695795; SN 754862; SN 745821). Three dams and reservoirs are linked by a series of tunnels and aqueducts to provide water for a hydroelectric power station adjacent to the lower, Cwm Rheidol, dam. Nant y Moch is the largest (SN 754862) and is a mass concrete buttress-type dam 170ft high and 1,150ft long. It impounds a reservoir of 5,700 million gallons at an elevation of over 1,100ft above sea level. Several aqueducts supply water to it from other streams in the catchment area. From Nant y Moch a 2.5-mile pressure tunnel 9ft 6in in diameter carries water to Dinas power station, whose tail race discharges into Dinas Reservoir (SN 745821).This has a spillway level 835ft above sea level and a capacity of 185 million gallons.

From Dinas reservoir a 2,100ft-long tunnel also 9ft 6in in diameter takes water to Cwm Rheidol power station, 160ft above sea level. This originally generated up to 42mW but was upgraded to 56mW when three additional turbines were added in 2004. The water discharges into Cwm Rheidol reservoir, which serves to regulate the flow of water downstream. All dams contain subsidiary power-generating sets producing electricity for local use.

The scheme was designed by Freeman Fox & Partners, James Williamson & Partners and Kennedy & Donkin and was built between 1959 and 1961 by Sir Alfred McAlpine Co. Ltd, Cementation Co. Ltd, Whatlings Ltd, and Taylor Woodrow Construction. There is a visitor centre at the power station, signed from the main road. (HEW 1239)

M5. Welshpool & Llanfair Light Railway (SJ 229072 to SJ 106068) was the first narrow-gauge railway built under the Light Railways Act of 1896. Running through the magnificent scenery of the Welsh border country between Welshpool and Llanfair Caereinion, it opened on 4 April 1903. Built to a 2ft 6in gauge, the railway originally started in the yard of the Cambrian Railways station at Welshpool where there were interchange sidings and locomotive and goods sheds. Passing through the streets of the town it emerged at Raven Square on the west side, the present terminus of the line. From here it has a steep climb, including a one-mile section at 1 in 29 to reach its highest point, Golfa Summit, 363ft above the starting point. It descends into the valley of the River Banwy, which it follows to Llanfair Caereinion station nine miles from Welshpool. There are four intermediate stations, Sylfaen, Castle Caereinion, Cyfronydd and Heniarth. As a true light railway the line is steeply graded and sharply curved, the curve at Dolrhyd Mill being only three chains (198ft) radius. There are few major structures; the two most notable are the six-arch masonry Brynelin Viaduct and a three-span steel girder bridge over the river. The engineer for the line was A.J. Collins, then chief engineer of the Cambrian Railways, and the contractor was John Strachan of Cardiff.

147 *'The Earl' at Cyfronydd Station. (© RC)*

Both passengers and freight were carried. Locomotive power was provided by two identical 0-6-0T tank engines supplied by Beyer Peacock in 1902, named *The Earl* and *The Countess* in honour of the Earl and Countess of Powys, owners of the nearby Powys Castle.

In 1923 the railway passed into the ownership of the Great Western Railway. The passenger service was discontinued in 1931 because of increasing competition from road vehicles. The freight service continued until 1956, the principal traffic being coal, agricultural implements, fertilisers and building materials westwards and timber, livestock and agricultural produce eastwards.

A Preservation Society was formed in 1956 and passenger services restarted in 1963, at first only from Llanfair Caereinion to Castle Caereinion, later extended to Sylfaen Station and in 1981 to a new terminus at Raven Square, Welshpool. The original locomotives are still in use together with an interesting collection of motive power and rolling stock from around the world. (HEW 1483)

M6. Montgomeryshire Canal (SJ 254203 to SO 116921) is a narrow 23½-mile-long branch from the Ellesmere Canal. This canal ran from an end-on junction with the Llanymynech Branch of the Ellesmere Canal via Welshpool to a terminus at Newtown. It was built in two stages – Carreghofa to Garthmyl in 1797 and Garthmyl to Newtown in 1819. The canal was abandoned but is being restored in stages. Complete restoration is planned except for the final three miles into Newtown. Engineers for the first section were John Dadford jr (to 1797), Thomas Dadford sr and Thomas Dadford jr. For the second stage the engineer was Josias Jessop with John Williams as resident engineer.

The original act of 1794 allowed for a canal from Porthywaen limestone quarries to Newtown, with branches to Llanymynech and Guilsfield. As actually built, a junction was made with the Llanymynech branch at Carreghofa, where a feeder from the River Tanat joined it and a tramway to the quarries. A new company was set up in 1815 for the western branch to complete the canal to Newtown. The eastern branch was transferred to the Shropshire Union in 1847 and the western branch three years later. The canal eventually came into the ownership of the London, Midland & Scottish Railway but was abandoned by them in 1944. Penarth Weir two miles downriver from Newtown supplies the water for the western branch of the canal. This used to be supplemented by a river-actuated pump at Newtown.

There are several aqueducts on the Welsh length of the canal; four aqueducts are described below. Some required rebuilding or strengthening within 20 years or so, rather more than other canals of the time. One possible reason is that the structures had to be founded in deep variable alluvial deposits in the flood plain, which were inadequate as a long-term foundation for the very heavy aqueducts. Aqueducts are heavily loaded structures, carrying not just the weight of water but a thick layer of dense puddled clay used to make the bottom and sides of the waterway water-tight. The foundations for the arches, probably timber piles capped with a horizontal timber grid, may not have been adequate for the weight of the aqueduct, resulting over time in differential settlement and distortion of the masonry. A masonry arch bridge just upriver of Vyrnwy Aqueduct has also been strengthened with tie bars, so the problem was not peculiar to the canal construction.

In addition there was pressure to construct the canal as soon as possible and the engineers concerned were also heavily committed to other canals being built at the same time. By the time G.W. Buck was appointed engineer in 1819 the use of cast iron had been developed and a lighter form of aqueduct could be constructed, which helped overcome most of the problems. (HEW 1205)

148 *Vyrnwy Aqueduct. (© RC)*

M6.1. Vyrnwy Aqueduct (SJ 254196). Built about 1794-7 by John Simpson and William Hazledine, two experienced contractors who later built Pontcysyllte Aqueduct, this five-span masonry arch aqueduct carrying the canal over the River Vyrnwy was designed by John Dadford and then Thomas Dadford after a partial collapse of one arch during construction. It has a length of 292ft and is 31ft wide. The arches are extensively supported by cast-iron rods three inches in diameter through and under the arches, connected to cast-iron T-beams pinned to the arch spandrels. This ironwork was added in 1823 as part of extensive repairs to stop the aqueduct splitting apart. Every arch was fractured and the aqueduct leaking.

The trough was originally brick-lined but was repaired in the 1980s with concrete. The aqueduct is supplemented by two sets of flood arches, one of four and the other of three arches, each side of the main aqueduct. The structure is listed Grade II*. Access is via the B4398 from Llanymynech or the B4393 from Four Crosses both on the A483. The towpath at this point is part of the Offa's Dyke Path. (HEW 1112)

M6.2. Welshpool Aqueduct (SO 227074). Built to carry the canal over the Ledan Brook, this masonry aqueduct was rebuilt in 1836 as a cast-iron and masonry aqueduct. It was designed by J.A. Sword, who took over the position of engineer from George Buck in 1833 when he left to become one of Robert Stephenson's engineers. The ironwork is believed to have been supplied by Plas Kynaston Foundry. The cast-iron trough is 28ft long, eight feet wide and 5ft 2in deep, built from a separate base and side plates bolted together. The trough is given additional support by two cast-iron I-beams set under the edges of the trough with transverse tie rods. The trough is flanked by two areas 6ft 4in wide supported by stone segmental arches, one of which is the towpath. At the north end of the aqueduct is a steel girder bridge which formerly carried the Welshpool & Llanfair narrow-gauge railway.

Listed Grade II, the aqueduct is immediately south of the recently reconstructed Canal Wharf where canal trips are available in the summer. There is a large car park and the Canal Wharf is signed from the main road. (HEW 1836)

149 *Welshpool Aqueduct. (© RC)*

M6.3 Brithdir Aqueduct (SJ 198022). Designed by George W. Buck, this cast-iron aqueduct was built in 1819 to replace an earlier masonry aqueduct taking the canal over the River Luggy. It has a clear span of 26ft 10in between abutments, with a total trough length of 38ft. The original span was greater but the abutments were extended about 1890 to remedy sagging of the trough. Three raked struts between the abutment and the first base section joint were probably added at the same time. Overall width of the trough is 12ft with an eight-foot waterway and a four-foot towpath on the south side. It is listed Grade II.

The aqueduct is 100m south of Brithdir Lock and can be accessed from the rear of the *Horseshoe Inn*, just north of the

150 *Berriew Aqueduct.* (© RC)

Berriew junction on the A483, or from an unclassified road along the south side of the inn to a gate into the Brithdir Nature Reserve. (HEW 0522)

M6.4. Berriew Aqueduct (SO 189007). Opened in 1797, this four-span brick aqueduct crossing the River Rhiw on the Eastern Branch was designed by John Dadford jr and built by John Holbrook of Oswestry. As a result of construction difficulties it was repaired by Thomas Dadfield sr before opening and was extensively rebuilt in 1889 when clad in the present blue brick. The two main segmental arches are 30ft span with an 8ft 4in rise. Subsidiary arches to the north and south over roadways are 10ft span with a 3ft 9in rise. The waterway is 8ft 6in wide with a total trough length of 147ft, flanked by an 11ft towpath on the south-east side and a grassed area on the north-west, giving an overall width of 33ft 3in. In 1985 a new concrete trough was constructed with blue brick coping and the canal was re-opened to navigation. Listed Grade II.

Access is via a minor road on the south side of the river, from the A483 at Refail to Berriew, which passes under the aqueduct, or from the B4390 to Berriew from the A483. Park at the canal over-bridge before the village and follow the towpath south for about 400m. (HEW 1465)

M6.5. Penarth Weir, Newtown (SO 140927). In 1819 the final section of the Montgomeryshire Canal was opened from its temporary terminus at Garthmyl to Newtown. In order to ensure an adequate supply of water to the upper end of the canal at Newtown, the Engineer, Josias Jessop, and his resident engineer, John Williams, constructed a weir across the River Severn below Newtown, from which a short feeder channel conveyed water diverted from the river into the canal.

The masonry weir has a total height of eight feet, in two steps. The upper crest is three feet wide, followed by a slope about ten feet long leading to an apron 11ft 10in wide; this slopes up to the lower crest, which is 3ft 6in wide, with a final slope down to the base of the weir. The weir is curved in plan, forming a segment of a circle of about 150ft radius. It is convex upstream and the crest length is about 140ft. On the south bank of the river the weir abuts a vertical rock face. On the north side there is a bypass sluice and a fish ladder, both of which are of relatively modern construction.

151 *Penarth Weir.* (© RC)

About 90ft above the weir is the entrance to the canal feeder, which runs north-eastward parallel to the river for about 350 yards to join the canal. The feeder enters a nine-foot-wide brick arch culvert leading to an open channel about 13ft wide at water level. A modern timber bridge carries the towpath across the mouth of the feeder. Access is down a minor road by the church to the sewage works, eastwards for 300m along the canal and then through the Nature Reserve to the weir a quarter of a mile away. (HEW 1279)

M7. Llandinam Bridge (SO 025886) was the first cast-iron bridge in Montgomeryshire. This arch road bridge over the River Severn was built in 1846. A previous bridge was described as 'out of repair' in 1707 and became a county bridge in 1822. It was cast by the Hawarden Ironworks to the requirements of Thomas Penson, County Surveyor, who also designed the bridges at Abermule and Caerhowel.

152 *Llandinam Bridge.*

It was constructed on the 'Telford' model and is similar to his bridges elsewhere in Britain. Penson had produced a cast-iron design with Telford in 1824 for Llanymynech Bridge on the border between the two counties. The design was not used but gave Penson an understanding of the design principles involved in the use of the material. Telford used William Hazledine's Plas Kynaston Foundry for his bridges and after Hazledine's death in the 1840s some of his castings and patterns were sold to the Hawarden Ironworks. Some of Hazledine's original castings or moulds may have been used here.

The single arch spans 90ft with a rise of nine feet and is made up of three ribs, each consisting of five segments; each segment in turn has five cruciform-section lattice panels. The spandrels are also of open cruciform pattern with four-inch cruciform members. The bridge is stiffened laterally by rectangular and circular cross-members connecting the arch ribs. It has a width of 10ft 5in and carries a minor road over the River Severn. It currently has a three-ton weight limit and is listed Grade II*. The stonework contractor was Edward Jones of Llandinam. (HEW 0850)

Aft the east end of the bridge, adjacent to the main road, there is a statue of David Davies, the eminent railway contractor, owner of the Ocean Coal Company and the promoter of Barry Docks, who was born in Llandinam in 1818.

M8. Abermule Bridge (SO 162951) was the second cast-iron bridge built in Montgomeryshire. This elegant bridge over the River Severn just to the north-east of Abermule, designed by Thomas Penson, County Surveyor, was cast by the Brymbo Company iron foundry in 1852. Masonry was by David Wylie of Shrewsbury. The single 110ft-span arch, with a rise of 12ft 3in, consists of five 2ft 9in deep ribs, each with seven bolted segments. The transverse stiffening of the arch is by rectangular diaphragms at the segmental joints, circular tie rods at segmental centres and horizontal lattice bracing over the outer three segments. Diagonal cruciform-section bars run between the two outer ribs on each side over the two segments nearest the abutments and also connect the inner three ribs of the third segment from the abutment. The abutments are of ashlar masonry with rock-faced blocks on the corners. The bridge deck is of cast-iron plates with a macadam road surface.

An inscription cast into the outer two arch ribs records that it was the second iron bridge in the county of Montgomery. The width of 21ft between the parapet rails includes a 17ft carriageway. Although still in use, the bridge is no longer on the main road as this now bypasses Abermule. It has three-ton weight limit. (HEW 0342)

Immediately adjacent and to the north of the bridge is a 16ft-span cast-iron bridge over the Montgomeryshire canal, also by Brymbo Ironfounders, built in 1853.

153 *Caerhowel Bridge.*

M9. Caerhowel Bridge (SO 197982) is two-span cast-iron bridge of 1858. Another of the series of cast-iron bridges over the River Severn in this area, it is situated near Caerhywel Hall, some 2¾ miles downstream from Abermule Bridge. Designed by Thomas Penson, jr, County Surveyor of Montgomeryshire, it was also cast and built by the Brymbo Company. It replaced an earlier Dredge-type suspension bridge that collapsed under a herd of cows and which itself had replaced a timber bridge. It has two segmental arch spans of 72ft 8in, each with five ribs, and each rib has five segments. The arch is strengthened laterally by rectangular diaphragms at the segmental joints with circular tie bars at the segmental sections. Diagonal bracing bars of cruciform cross-section link adjacent ribs. The 21ft 9in deck is formed from cast-iron plates. The abutments and piers are of ashlar masonry. The bridge deck was reconstructed as a single lane with two footpaths to full highway loading in 2004. (HEW 0851)

M10. Talerddig Cutting (SN 931995 to SH 930001) is a 120ft-deep rock cutting, one of the deepest on a main line. The Newtown & Machynlleth Railway was built under an Act of 1856. Its route between the two towns of its name leaves the valley of the River Severn at Caersws and climbs over the hills to descend into the valley of the River Dovey at Cemmaes Road.

154 *Talerddig Cutting. (© RC)*

At Talerddig the single line reaches its summit level of 693ft above sea level. Here the original plan was for the line to tunnel through a steep rock outcrop, but the plans were changed and a deep cutting was excavated instead. Talerddig cutting is about 120ft deep at its maximum and is one of the deepest on British railways. The stone excavated from the cutting was used in the construction of the masonry structures on the line. To the north-west of the cutting the railway is crossed by a minor road on a concrete beam bridge.

The engineers for the railway were Benjamin and Robert Piercy and the contractors were David Davies of Llandinam and Thomas Savin, one of several railway and building contracts they built as partners before they parted company a few years later. It was opened in January 1863. (HEW 1829)

M11. Knuclas (Cnwclas) Viaduct (SO 250742) is a heavily castellated railway viaduct. In the late 1850s and 1860s a series of railways was proposed to link the lines of South Wales with the West Midlands. Part of what became known as the Central Wales Line was the Central Wales Railway running from Knighton to Llandrindod Wells and linking with the Knighton Railway (Craven Arms to Knighton). It was incorporated in 1859 and the line was built by the company but worked by the LNWR from its opening.

The Cnwclas Viaduct carries the single line of the railway over a side valley of the River Teme as the line starts its climb to the summit at Llangynllo Tunnel. The 470ft-long viaduct, designed by Henry Robertson and built by Richard Hattersley, is an elegant structure. There are 13 arches of 30ft span, the maximum height above the valley being 75ft. Semi-circular stone arches surmount stone piers and the whole viaduct is richly decorated with castellated stone towers at each end and a battlemented parapet. It is listed Grade II.

The section of the Central Wales line upon which the viaduct stands was opened on 10 October 1863. (HEW 0594)

155 *Knuclas Viaduct.*

M12. Elan Valley Reservoirs

In 1892 the Birmingham Corporation Water Act authorised the construction of reservoirs in the Elan and Claerwen Valleys south-west of Rhayader. James Mansergh was the Engineer. The initial works in the Elan Valley comprised three reservoirs, which were built by direct labour and finished in 1904. There is a visitor centre at Caban Coch and scenic drives around the reservoirs.

Caban Coch (SN 925646) is the first dam up the valley from Rhayader and is 610ft long, 122ft high and five feet wide at the crest. It is built of cyclopean mass concrete, which includes rubble 'plums' of up to 10 tons, and is faced with block-in-course masonry. The downstream face has an inwardly curved batter struck to a radius of 340ft to within 15ft of the top from which point the curvature is reversed. The area of the reservoir is 500 acres and the impounded capacity 7,815 million gallons. There is an interpretation centre close by giving information on the works. (HEW 0550)

156 *Garreg Ddu submerged dam with roadway over.*

A novel feature of the scheme is the submerged dam built across the reservoir at **Garreg Ddu** (SN 910640) about 1½ miles upstream of Caban Coch. When the water level is low this keeps it at the required level to feed the aqueduct while the water in the reservoir below can be

157 *Claerwen Dam.*

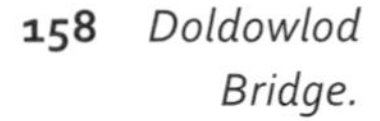

158 *Doldowlod Bridge.*

used to provide compensation water to the river. The intake for the aqueduct is from a valve tower just above Garreg Ddu and the aqueduct goes directly into a tunnel at that point.

The aqueduct (HEW 1194, SN 932652 to SP 003804) is over 73 miles long, taking water under gravity to a reservoir at Frankley just outside Birmingham at an overall gradient of about 1 in 2,300, a masterpiece of surveying and construction to achieve this very flat fall. It was designed to carry up to 75 million gallons a day.

There are two other reservoirs in the valley further upstream, **Pen-yr-Garreg** (SN 911675) and **Craig Goch** (SN 89468). In the neighbouring valley of the Afon Claerwen is a further reservoir. Here is the **Claerwen Dam** (SN 868635), which was built in 1946-52 by Edmund Nuttall, Sons and Co. and designed for the Birmingham Corporation by Sir William Halcrow & Partners. It is a curved mass concrete dam 1,066ft long and 180ft high above the river bed, with masonry facing on the downstream and upper portion of the upstream faces. The reservoir covers 650 acres and holds 10,625 million gallons. The central spillway is 540ft long, discharging down a stepped face into the stilling pool for appearance and also to slow the flow. (HEW 0186)

M13. Doldowlod (Ystrad) Bridge (SO 003617) is a privately owned suspension bridge over the River Wye about five miles south of Rhayader, within the Doldowlod estate of the Gibson-Watt family. The suspension chains span 120ft between 14ft 6in high cast-iron towers, each of which is constructed from five separate castings bolted together. Each leg has an upper and lower section with a decorative cross-member. The links of the chains consist of ¾in-diameter iron rods 7ft 4in long, the number of rods in each link varying from five at the towers to two at mid-span. The links are connected by one-inch diameter bolts. Inclined hangers attached to the chains at each link connection support a timber deck, an arrangement similar to that used by James Dredge for his suspension bridges such as Victoria Bridge, Bath and the one over the Kennet & Avon Canal at Stowell Park. The bridge was built about 1880, the exact date being uncertain. The ironwork was cast at Llanidloes Railway Foundry.

Access, by permission of the owner, is via a well-hidden private lane just south of Doldowlod Hall entrance. There is a small layby opposite and the bridge is about half a mile down the access. (HEW 1245)

159 *Cynghordy Viaduct.*

M14. Cynghordy Viaduct (SN 808418). The Central Wales Extension Railway from Llandrindod Wells to Llandovery was opened in 1868. This line included a number of interesting engineering works, this viaduct being the most spectacular. It has 18 spans of 36ft 6in, is curved in plan and 93ft high. It is simple in detail but the high quality of the coursed rock masonry and the six-ring brick arches, together with the rural setting, make it an impressive structure. It was designed by Henry Robertson and Charles Dean. The line is still in use.

The viaduct is about one mile north of the village of Cynghordy and can be reached from the A483 from

Llandovery. On the eastern side of Cynghordy take a left turn (northwards) down a lane and at the end turn right and carry on along a narrow lane to the viaduct. The viaduct is listed Grade II*. (HEW 0272)

160 *Spillway and Llyn Brianne Dam.*

M15. Llyn Brianne Dam (SN 791485) is the highest dam in the UK. Llyn Brianne Dam and Reservoir lie in mountainous country to the west of Llanwrtyd Wells and north of Llandovery. The dam was built to regulate the water supply in the River Towy (Tywi), from which there is a take-off 40 miles downstream to a treatment works to supply Swansea, Neath and Port Talbot. When at full capacity water is discharged through an upward-facing jet which as well as serving to oxygenate gives a spectacular display. Designed by Binnie & Partners, it was constructed by George Wimpey & Co. Ltd and the scheme was completed in 1972.

It is 900ft long at the crest and 30ft wide, and the height of 300ft makes it the highest dam in Britain. It is constructed of rock filling with a clay core. The dam design allows for a future raising of the crest by another 60ft. There is parking provision at the dam and a scenic drive along the reservoir. In 1997 a hydroelectric power station was added below the dam, designed by Hyder consultancy, then part of Welsh Water. (HEW 0552)

M16. Dolauhirion Bridge (SN 762361) was built in 1773. William Edwards, his two sons and his grandson, built a number of bridges in addition to his world-famous bridge at Pontypridd. Some of these, like Pontypridd, were single-span bridges with openings in the haunches and the best example is this bridge at Dolauhirion. Still in daily use, it spans the River Tywi about a mile north of Llandovery and just off the road to Rhandirmwyn and Llyn Brianne.

According to Jervoise, this bridge is the finest over the upper part of the Tywi. It is a single segmental arch with a span of 84ft and with a road width of 12ft. The circular openings in the sides are a distinctive feature of Edwards' bridges, developed from his experience of his failed first attempts at Pontypridd. The openings relieve the weight of the spandrels and abutments, reducing the upward thrust at the centre of the arch, and are not for flood relief. The bridge is a scheduled ancient monument and listed Grade I. (HEW 0167)

161 *Dolauhirion Bridge. (© Society for the Protction of Ancient Buildings*

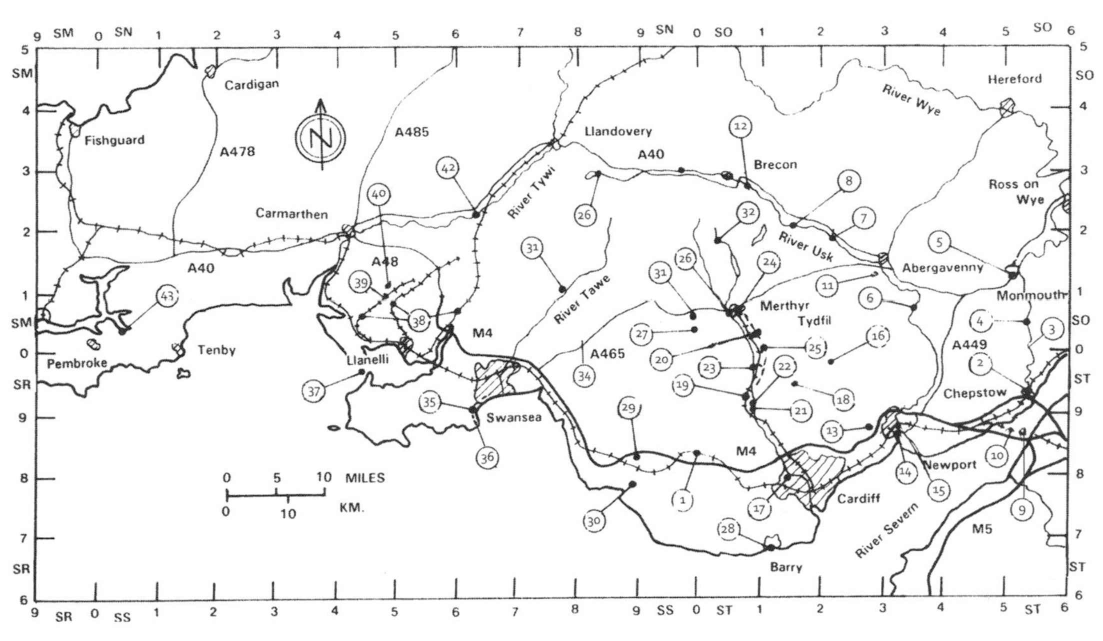

South Wales map.

S1. South Wales Railway
S2. Chepstow Bridge
S3. Wireworks Bridge, Tintern
S4. Bigsweir Bridge
S5. Monmouth Bridge
S6. Pant-y-Goitre Bridge
S7. Crickhowell Bridge
S8. Llangynider Bridge
S9 . B & SW Union Railway
S10. Severn Tunnel
S11. Mon. & Brec. Canal
S12. Brynich Aqueduct
S13. Fourteen Locks, Rogerstone
S14. Newport Transporter Bridge
S15. George Street Bridge, Newport
S16. Crumlin Viaduct
S17. Melingriffith water pump
S18. Hengoed Viaduct
S19. Taff Vale Railway
S20. Quakers Yard Viaduct
S21. William Edwards Bridge, Pontypridd
S22. Berw Road Bridge, Pontypridd
S23. Pont-y-Gwaith Bridge
S24. Pont-y-Cafnau Bridge
S25. Merthyr (Penydarren) Tramroad
S26. Cefn Coed-y-Cymmer Viaduct
S27. Robertstown Bridge, Aberdare
S28. Barry Docks
S29. Glanrhyd Bridge
S30. New Inn Bridge, Merthyr Mawr
S31. Nant Hir Reservoir
S32. Upper Neuadd Dam
S33. Usk Dam, Sennybridge
S34. Neath and Tennant Canals
S35. Swansea & Mumbles Railway
S36. Mumbles Pier
S37. Whiteford Point Lighthouse
S38. Llanelly Railways
S39. Gwendraeth Fawr Bridge
S40. Pont Spwdwr, Llandyry
S41. Twrch Aqueduct, Ystradgynlais
S42. Llandeilo Bridge
S43. Carew Tide Mill

The industrial history of South Wales has been very much influenced by its geology and geography. Underneath most of the area is a bowl-shaped coalfield with associated ironstone and limestone deposits, outcropping in the north near the Brecon Beacons and the Black Mountains and in the south on the northern edge of the Vale of Glamorgan. The eastern edge is broadly a line north-south above Newport and the western edge is west of Llanelli. There is a small outcrop in South Pembrokeshire. In the east the coal deposits are bituminous, in the centre high-quality steam coal and in the west anthracite. The extreme west and east of the region is predominantly agricultural.

In the 18th and 19th centuries ironworks grew up along the northern edge of the basin and to a lesser extent the southern and western ends to exploit the ironstone and limestone using the plentiful supplies of timber and water for power. Later as the woodlands were cut down the abundance of coal provided a further boost to the industrial development. In the 1830s a method of smelting iron using anthracite was developed, further expanding the iron industry and the coal mining in the west.

Pack-horse routes to take the iron products to the ports followed as the industry grew by canals and tramroads. Steam began to replace horse-power in the 1840s and improved pumping methods enabled deeper pits to be sunk nearer the centre of the coalfield basin, tapping huge amounts of high-quality steam coal to fuel the rapid world-wide expansion of steam power for locomotives, shipping and industry. Coal exports to all parts of the world required the construction and the expansion of ports and harbours. The continued demand for iron and, later, steel products meant that larger iron and steelworks were needed. The demand for iron ore as the ironstone seams were worked out also required the development of import facilities.

162 *Richard Trevithick (1771-1833). (© ICE)*

163 *Isambard Kingdom Brunel (1806-59). (© ICE)*

All this was not achieved without great effort because of the geography in the south and east; a series of steep narrow valleys running from the north to the coast presented major challenges to the early civil engineers. Canals sloped steeply to the sea and had locks considerably deeper and more frequent than was the practice in England, while the tramways and railways needed many viaducts across the narrow valleys and tunnels between them. The associated rapid growth in population created further demands.

The coal trade began to decline in the 20th century with the advent of petroleum products for power. The steel trade continued but export markets declined in the 1950s and 1960s as countries developed their own heavy industries. Two large steelworks were built on the coast to replace several older and less efficient ones. Major industrial complexes concentrated on the seaboard. The main emphasis for communication changed to an east-west one and, as the coal handling facilities in the docks became redundant, the smaller docks and ports closed and the larger ones developed these areas for general cargo and for industrial and business parks. Milford Haven developed as a major UK deep-water facility for the import of oil and gas.

Following the closure of the deep mines much of the infrastructure has disappeared, land reclamation operations have created new areas for small industrial businesses, and the old rail tracks have been redeveloped as new or improved highways. Many of the canal and rail structures have been demolished or buried under new construction but some notable examples of the work of early civil engineers remain. The new highways themselves have also had to meet the same challenges of the topography and some fine examples of modern civil engineering have been produced.

S1. South Wales Railway (ST 536937 to SM 967048) is a major early railway with I.K. Brunel as engineer. The first prospectus for the South Wales Railway was issued in 1844 and an Act passed in 1845 to build a line from the Gloucester branch of the Great Western Railway (GWR) at Grange Court to Pembrokeshire to form a link to Ireland. Brunel designed it to his preferred broad gauge of seven feet. Terminals at Fishguard, Pembroke Dock and Abermawr were examined but eventually were rejected on cost grounds, although work actually started towards Abermawr. Neyland in Milford Haven was chosen instead.

164 *Chepstow tubular suspension bridge. (© SKJ Collection)*

In 1850, 75 miles of the line were opened from Chepstow to Swansea, the bridge at Chepstow was completed in 1852 and the link to Grange Court, near Gloucester, was made.

The Swansea to Carmarthen section opened in 1852. In 1853 the section between Carmarthen and Haverfordwest was opened and the final section to Neyland in 1856 completed the route. The SWR had strong links with the GWR and eventually became part of it in 1863. Apart from the bridge over the Wye at Chepstow, notable structures on the South Wales Railway were the crossing of the Usk at Newport (where the timber bridge caught fire before the line opened and was hurriedly replaced with a wrought-iron bowstring girder bridge, subsequently rebuilt and widened); Landore Viaduct, on which many timber spans were replaced by iron in 1888 and by steel in 1979; Llwchwr (Loughor) Viaduct, another timber viaduct whose deck is now in steel but with Brunel's timberwork still existing at the lower level; and a bridge with a drawbridge opening span at Carmarthen, built by Brunel in 1854 and rebuilt in 1910 as a Scherzer rolling lift bascule bridge. Models of the main span of the original Landore Viaduct are in the Swansea Museum and the National Railway Museum in York.

The curves at Neath are not conducive to high-speed running and the few stiff gradients thereabouts made heavy demands on motive power. This situation was eased in 1913 by the construction of a bypass line which crossed the River Neath by an interesting swing bridge, originally built by the Rhondda & Swansea Bay Railway. Five fixed spans of plate girder construction, varying in length from 40ft to 52ft 6in, have been renewed. The movable span is a Pratt truss with a curved upper boom, swinging about its centre and resting on a cast-iron roller race, the span being 167ft 6in. Originally operated hydraulically, the bridge is now fixed. It is the only opening bridge of this type in Britain built both on the skew and on a curve. Finch of Chepstow supplied the steelwork, Sir William Armstrong of Newcastle the hydraulic machinery.

At Llansamlet, between Neath and Swansea, four 70ft-span stone arches, which may be compared with the Chorley flying arches, spring from the sides of a cutting. Brunel seems to have built them to permit steeper cutting slopes, reducing the risk of slippage and so saving on excavation.

165 *Carmarthen Railway Bridge.*

Bridgend station was overhauled in 1980 and is now a pleasant mixture of the original Brunel buildings and modern railway architecture, which received awards from HRH the Prince of Wales and from the Development Corporation of Wales. Chepstow Station is very much original and the 1850 train shed to the west still exists as an industrial building. The building was an engine shed until the final link to Gloucester was available and then became a goods shed.

The continued use of the broad gauge by the GWR after it had been officially declared non-standard in the United Kingdom was unfortunate since the huge output from the South Wales collieries demanded mineral wagons which would be capable of working over any part of the English rail network irrespective of track gauge as well as up the Welsh valleys themselves. Even Brunel had thought in 1840 that something smaller than the broad gauge was desirable when he built the Taff Vale Railway where the topography favoured a narrower gauge. His other lines in South Wales, however, he still built to broad gauge. By 1866 so great was the clamour in South Wales to convert the main land route eastward into England to standard gauge that this was accepted by the GWR and by 1872 its broad gauge lines had been converted.

166 *Neath River Swing Bridge. (© TW)*

167 *Llansamlet Arches. (© Owen Gibbs Collection)*

168 *Bridgend Station in 1850 from a lithograph by Newman & Co. showing a broad gauge train and the station building which still exists. (© National Library of Wales)*

The long detour via Gloucester was still a disadvantage for London, Bristol and West of England traffic. Brunel's original proposals had envisaged a crossing of the Severn at Hock Cliff (SO 730090) but this was vetoed by the Admiralty. Brunel also planned a link between Bristol and South Wales in 1846 with a ferry at New Passage and this was constructed in 1857 by the **Bristol & South Wales Union Railway (S9)**. However, it was not until 1886 that the GWR resolved the situation with the construction of the **Severn Tunnel (S10)** and, for London traffic, by a direct link from it to Swindon in 1903. (HEW 1199)

169 *A section of Brunel's supporting tube from Chepstow is preserved at the entrance to Fairfield's Works near the bridge.*

South-East Wales

The Bridges in the Wye Valley (S2-S5)

There are a series of interesting bridges in the Wye Valley between Chepstow and Monmouth. The lowermost bridge crossing the Wye is the cable-stayed bridge on the M48 motorway opened in 1966. The Welsh/English boundary is the River Wye and this bridge is somewhat overshadowed by its famous neighbour on the English side, the Grade-I listed Severn Suspension Bridge, to which it is linked by a short viaduct. Upstream of the Wye Bridge there is a series of bridges over the river, some of which are described below.

S2. Chepstow Bridge (ST 536943) is a unique type of multi-span cast-iron bridge. To the historically minded, mention of Chepstow brings to mind the medieval castle and Brunel's railway bridge (now replaced). However, just upstream, and not far from Chepstow Castle, is an older and interesting five-span cast-iron bridge carrying the Chepstow to Tutshill road over the River Wye, which opened in 1816 to replace an earlier timber bridge (shown in J.M.W. Turner's painting of the castle). The old bridge was frequently damaged and eventually the Gloucester Justices decided to replace it. Watkin George of Cyfarthfa and John Rennie each prepared designs for cast-iron spans on the existing piers in 1811-12 and Rennie also submitted a design for a 250ft span cast-iron arch. However none were accepted and in 1814 a contract was let to John Urpeth Rastrick, Robert Hazledine, Thomas Davies and Alexander Brodie of Bridgnorth. Some of the ironwork was cast at Calcutts, located beside the River Severn in Shropshire. Calcutts at that time was owned by Brodie and later taken over by William Hazledine. Rastrick was one of the most important engineers of the time and was engineer at the Bridgnorth Foundry until 1817, becoming a major partner in 1819 of Foster, Rastrick and Partners and was involved in many early railway schemes.

170 *Chepstow Bridge.*

The centre span is 112ft, flanked by two spans of 70ft and two end spans of 34ft on each side. The overall length is about 490ft and width 20ft. The spandrel fillings are of the radial grid pattern, not unlike those on some of Telford's bridges. The graduated spans, vertically curved road profile, parapet fence and decorative lamp standards combine to give the bridge its attractive appearance. The bridge uses abutments partly from the original bridge.

171 *Wireworks Bridge.*

In spite of its age, the bridge has continued to carry traffic with the help of traffic signals which limit the flow to one lane alternately in each direction. The centre span was strengthened in 1889 by the addition of steel ribs and the abutments by sheet steel piling; further extensive repairs were carried out in 1979-80. The bridge is listed Grade I. (HEW 0145)

S3. Wireworks Bridge Tintern (SO 530003) is a three-span wrought-iron railway bridge now used as a public footpath. The bridge was completed in 1872-5. It was designed by S.H. Yockney & Sons of Westminster for the Wireworks Branch of the Monmouth & Wye Valley Railway. The contractors were Reed Bros. of London, with ironwork supplied by the Isca Foundry of Newport. It was designed to carry no more than a locomotive and three trucks. The works, founded in 1556, changed to tinplate production in 1880 and closed in 1901 but the line continued in use as a horse-drawn tramway until 1935.

It is a wrought-iron lattice girder through-bridge with three equal spans of 66ft 6in. The girders are 6ft 6in high at 13ft centres with nine diagonally braced panels 7ft 6in long in each. Transverse and diagonal ties support the timber deck. The Welsh end is supported by a sandstone ashlar abutment; the English end has two small masonry side spans. It was listed Grade II in 2002. (HEW 0930)

S4. Bigsweir Bridge (SO 539051) is a large-span cast-iron bridge with an unusual arrangement of spandrel bracing. Travellers on the A466 between Chepstow and

172 *Bigsweir Bridge from the Welsh side.*

Monmouth cross from Wales into England over this bridge, crossing back a few miles further north at Redbrook. It spans the River Wye and blends harmoniously into its beautiful scenic setting. The bridge formed part of a toll road between St Arvans and Redbrook authorised in 1824, and the bridge was completed in 1828. The designer was Charles Hollis of London and the contractors Bough & Smith.

The elegant cast-iron arch of 164ft span has four ribs of dumb-bell section in 16 segments with plain 'N'-spandrel bracing. It is surmounted by a cast-iron balustrade. The arch segments, cast at Merthyr Tydfil, are particularly well made.

Around the mid-19th century two masonry-arched flood spans were added at each end, making the overall length 321ft. The mock string course cast on the main arch at deck level is continued in masonry over the faces of the flood spans.

Despite its age, and a carriageway only 12ft wide, the bridge copes with the heavy tourist traffic along the Wye valley with the aid of traffic signals and a 17-tonne weight restriction. The bridge is listed Grade II*. The old tollhouse in a dilapidated condition can still be seen at the Welsh end of the bridge. (HEW 0361)

S5.1. Monmouth Bridge (SO 505125) is the only remaining bridge in Britain with a fortified gateway on the bridge itself. The Monnow Bridge at Monmouth was built in the 13th century and is a scheduled ancient monument listed Grade I. In 1988 work in the river revealed timbers from an earlier bridge which has been dated to 1140. It is unique in having its fortified tower on the bridge. A bridge at Warkworth, Northumberland (HEW 0696), has a fortified tower at its southern end but not on the bridge itself.

The bridge, totalling 114ft in length, has three semi-circular masonry arches, each having three wide ribs. It has been widened by some 3ft 6in on the upstream side and 5ft on the downstream side to give a width between parapets of 24ft. The widening has been carried out with great skill so that it blends well with the original structure.

The fortification is arched over the roadway above the eastern pier of the bridge. Surmounting the archway there is a room 36ft long by 10ft wide covered by a pitched roof. The bridge is now pedestrianised, a new road bridge having been built just downstream. (HEW 0146)

S5.2. Inglis Bridge, Monmouth (SO 508135) is the only remaining example of a First World War temporary bridging system in public use. It was constructed in units that could be manhandled under field conditions. Developed in 1915-16 for use in combat

173 *Monmouth Bridge.*

174 *The Inglis Bridge at Monmouth.*

conditions, it was designed by Professor Sir Charles Inglis while he was serving as an officer in the Royal Engineers. It has few component parts and can be constructed using the minimum of mechanical aids since all the components can be manhandled into position. The tubular sections are pinned into cast sockets to make 15ft bays. A maximum of six bays could be built and greater load capacity could be achieved by doubling or trebling the Warren girders. It was designed for rapid erection and deployment and a 60ft span could be built and launched by a troop of 48 men in less than 12 hours.

It was superseded early in the Second World War by the better known Bailey Bridge System which has been used extensively for both military and civil temporary bridging, an example of which can be seen outside the Army Museum in the town.

This bridge was erected in 1931 by the Royal Monmouthshire Royal Engineers (Militia) to replace an earlier timber bridge. It can be reached from the Osbaston road just off the A466 at the north of the town and is the access to the public recreation fields and the army camp.

S6 Pant-y-Goitre Bridge (SO 348089) is a three-arch masonry bridge of unusual design. It was built in 1824-6 as a design-and-build contract by John Upton of Gloucestershire. It carries the Llanvihangel Gobion to Usk road over the River Usk. The bridge has a centre span of 58ft and side spans of 39ft. The arches are semi-elliptical in form. The spandrel faces are pierced by cylindrical voids close to the springings and between the springings there is a larger void, nine feet in diameter, over each pier. Similar nine-foot diameter voids through each abutment serve for flood relief. An unusual feature is the way the masonry in the spandrel faces is laid with the joints radial to the arch rather than horizontal. Each wing wall curves outwards and is terminated in a semi-cylindrical pilaster rising from foundation level. This is a most attractive bridge, graceful in shape and in beautiful surroundings, and is Grade II* listed.

There is a plaque on the nearby bridge at Llanellen (SO 305110) stating 'Designed and built by John Upton of Gloucester 1821 for the County of Monmouth'. In fact, it

175 *Pant-y-Goitre Bridge.*

was built in 1823-6 by Upton but to the design of Thomas Waters, the County Surveyor. Upton, who was the Surveyor of the Old Stratford & Dunchurch Turnpike Trust, appears to have been a bit of a rogue who skipped bail when prosecuted for fraud when in charge of the Holyhead road in Northampton, and later left for Russia to avoid his creditors. He died in 1853. (HEW 0704)

176 *Crickhowell Bridge.*

S7. Crickhowell Bridge (SO 214181). Just south of the A40 between Crickhowell and Bwlch two rather splendid 16th- or 17th-century masonry bridges cross the River Usk. That at Crickhowell, which is a scheduled ancient monument and listed Grade I, has 12 spans on the upstream face and 13 on the downstream face, the largest of the spans at the northern end opening out to two exits. The flat segmental arches range in span from 16ft 10in to 39ft 6in. The bridge is built of coursed masonry with cutwaters extending up to the top of the parapet level to form pedestrian refuges. It is noted in the 1690 Quarter Session records as 'ancient'. It was rebuilt by a local stonemason, William Powell of Llangattock, under a contract let in 1706.

It was widened on the downstream side in 1809-10 by Benjamin James of Llangattock, another local stonemason, who later became County Surveyor of Monmouthshire, but it can still only carry single-line traffic. Its size and its position on the A4077, which turns south-west off the A40 through the village, suggest that the main coaching road once crossed it to take a route close to that adopted for the canal between Brecon and Abergavenny. (HEW 1223)

S8. Llangynidr Bridge (SO 151202). Five miles upriver of Crickhowell, Llangynidr Bridge carries the B4560 road, about ½ mile south of its junction with the A40. The six segmental arches vary in span from 22ft to 30ft 6in. The arch rings have two courses of voussoirs, the outer course being shallower than the inner but overhanging it somewhat. The roadway is only eight feet wide.

The bridge stands in a well-frequented beauty spot and though shorter than Crickhowell Bridge it is the more impressive of the two, with its massive piers and its arches standing higher above the river. It is Grade-I listed.

177 *Llangynidr Bridge.*

The contractor here was the same William Powell who rebuilt Crickhowell, in this case under a contract let in 1700, and like Crickhowell the bridge has triangular cutwaters to the piers, carried upwards to the tops of the parapets to provide pedestrian refuges. It is described in the Quarter Sessions records as having been repaired in 1707. (HEW 1247)

S9. Bristol & South Wales Union Railway (ST 609726 to ST 504882) was a major step in reducing the distance by rail in South Wales. Brunel had proposed several alternative bridge crossings of the River Severn below Gloucester, but none had been acceptable to the Admiralty and the townspeople of Gloucester who were concerned for their trade. Hence his eventual choice of this rail and ferry link, crossing at New Passage.

Opened in 1863 for passengers only, these 11 miles of broad-gauge railway shortened the distance between Bristol and Cardiff from 94 to 38 miles. Originally planned by Brunel, the final route was engineered by R.P. Brereton, who had been Brunel's principal assistant. The line from Bristol led to New Passage at Redwick on the Severn (ST 543863). Here the trains ran onto a 1,635ft-long timber jetty and a steam ferry crossed the river to a similar jetty on the other side at Black Rock (ST 516881), from whence a mile of track connected with the Gloucester to Newport line of the South Wales Railway at Portskewett. The jetty heads incorporated stairs, ramps and floating pontoons.

Earlier crossings by ferry used Old Passage from Aust (ST563891) to Beachley (ST 557908), a route that later was used by a car ferry which operated until the opening of the Severn Bridge in 1966. The Severn Bridge Railway Company succeeded in gaining approval for a bridge just above Sharpness and this was opened in 1879. It provided a link for the Midland Railway to Bristol. It was a single-track crossing and remained in use until 1960 when it was severely damaged by a barge. It was demolished in 1969.

When the Severn Tunnel opened in 1886 the ferry was discontinued but much of the line on the Bristol side was incorporated into the new route. At Black Rock on the north bank a few pile stumps remain and just back from the foreshore are some masonry remains of an arch carrying a footpath over the rail cutting. (HEW 1033)

S10. Severn Tunnel (ST 480876 to ST 545854) was a major Victorian engineering feat and until recently the longest tunnel in Britain. The construction of the 4 mile 624 yard-long tunnel further shortened the route between London and South Wales and is a story of engineering triumph over water and tides. The contractor was T.A. Walker and the Engineer Sir John Hawkshaw.

The tunnel crosses the River Severn between Pilning and Sudbrook where the river is 2¼ miles wide with a tidal range of up to 50ft. The main river channel, the Shoots, is 80ft deep below the general bed level. The Great Western Railway, with Charles Richardson as Engineer, began work in 1873 with the sinking of a shaft at Sudbrook on the Welsh side, and the driving of a drainage heading towards the river. Progress was only just under a mile 4½ years later. In 1877 contracts were let for additional shafts on both sides and headings on the line of the tunnel. In 1879 when the headings had almost met there was a severe inundation of fresh water from an underground spring on the Welsh side flooding the Sudbrook workings. A contract was then let to Walker for the completion of the works but it was not until January 1881 that the inflow, called the Great Spring, was overcome and the works could be dewatered.

178 *The Welsh Portal. (© RC)*

In April 1881 water broke into the workings on the Gloucestershire side from the sea bed. This was sealed with clay and concrete. In October 1883 the Great Spring broke through again and a week later a very high tide flooded all the workings from the deep cuttings on each side. A heading was driven from Sudbrook to intercept the Great Spring and it was contained in a short section of tunnel below the main one, enabling the main drive finally to be dewatered.

The tunnel has a brick lining up to three feet thick and the full length was completed in April 1885. In September a special train took a party including Sir Daniel Gooch, Chairman of GWR, through the tunnel and it was formally opened to goods traffic on 1 September and to passenger traffic three months later.

A shaft was sunk alongside the tunnel at Sudbrook and six very large Cornish beam engines were installed to pump out up to 20 million gallons of water a day from the Great Spring. These have now been replaced by electric pumps. The water continues to flow at the same rate, some is discharged into the river and some added to the water supply network in the area. (HEW 0232)

179 *Crumlin branch at Allt-yr-yn.*

S11. Monmouthshire & Brecon Canal (ST 310890 to SO 045285) is a well preserved early canal. This canal is the amalgamation of the original Monmouthshire Canal from Newport to Pontnewydd (with a western branch to Crumlin), which opened fully in 1799 with a short extension south in 1814, and the Brecknock & Abergavenny Canal, built between 1796 and 1799 and connected to the Monmouthshire Canal at Pontymoile in 1812. The latter bought out the former in 1865 to form the Monmouthshire & Brecon Canal Company. The company was sold to the Great Western Railway in 1880.

The Monmouthshire Canal was built with tramroad links up the Sirhowy Valley and others to serve the numerous ironworks, limestone quarries and collieries in the area as far as Ebbw Vale and Blaenavon, as well as to the developing area of Pontypool and Pontnewydd where the canal terminated. Thomas Dadford junior, then working on the Leominster and Glamorganshire Canals, was appointed as Engineer. The first section was completed on the western branch to Crumlin in 1794, the main line, the eastern branch, between Newport and Pontnewydd opened in 1796. Congestion at the southern terminus in Newport was relieved by an extension southwards in 1814.

180 *Canal head at Brecon.*

The Brecknock & Abergavenny Canal in 1793 was planned to run from the River Usk just above Caerleon to Abergavenny and then to Brecon. An alternative to link to the Monmouthshire Canal at Pontymoile, near Pontypool, was proposed instead and this was surveyed by Dadford. Tramroads up the Clydach Valley from Gilwern to Blaenavon and Beaufort were included in the Act, providing shorter routes to access the Monmouthshire Canal system than down the Ebbw valley to Crumlin. Dadford had been

181 *(Above) Canal at Talybont.*

182 *(Right) Brynich Aqueduct.*

engaged as Engineer in 1795 and construction began around 1797. The length from Gilwern to Brecon was completed in 1800. The canal served river wharves at Gilwern just below Abergavenny, transporting coal and other products to Brecon, one of the purposes of its construction. The second purpose of the canal, the connection south to the Monmouthshire Canal, remained incomplete until 1812 when pressure from the growing trade in iron products required extra capacity.

Much of the main line of the canal is now in use as a recreational facility. Although it closed to traffic about 1933 it remained as a water supply to industry at the southern end, ensuring its continued existence until the increased interest in leisure facilities in the latter part of the 20th century. The canal head at Brecon has been completely restored and most of the towpath is accessible for walking and cycling. Boat hire and boat trips are available. The Crumlin branch is not yet fully restored but the towpath is also a walking and cycling route.

S12. Brynich Aqueduct (SO 079273) is a four-span segmental arch masonry structure in the Brindley tradition carrying the Brecknock & Abergavenny Canal (later the Monmouthshire & Brecon Canal) over the River Usk, just south of the A40 road, about two miles east of Brecon and is accessible from the B4558 to Llanfrynach. It was built in 1799-1800, the engineer being Thomas Dadford, junior. The contractors were Benjamin James and Walter Walters.

It is approximately 210ft in length. The canal width is 12ft within an overall width of 31ft, and the water level of the canal is about 35ft above river bed level. The spans are approximately 36ft and the pier widths 7ft 6in. (HEW 0337)

S13. Fourteen Locks, Rogerstone (ST 286884 to ST 279886) is the most spectacular set of locks in South Wales, near Newport on the Crumlin branch of the Monmouthshire Canal, north of the M4 motorway between junctions 26 and 27. There are 14 locks in all, only the deep lock chambers now remaining. A series of side and header ponds to conserve water can also be seen. The short lengths between adjacent locks are connected to side pounds and there are several header pounds between groups of locks in the flight. Thus water released on the downward passage of a boat could be partially reused on later upward passages. The upper part of one lock chamber is widened for reasons not known but possibly for a weighing station.

In all, the series of locks gave a rise of 168ft in approximately half a mile of travel, giving an average rise of 12ft per lock. Thomas Dadford junior was the engineer and contractor for the canal and the Crumlin branch was fully completed in 1798.

The uppermost lock and basin have been fully restored and the masonry of the other lock chambers has also been restored. At the head of the flight is a small Canal Museum and Interpretative Centre run by the Canal Trust, which is signed from the main road at High Cross. The towpath is now part of National Cycle Route 47 and well maintained, and the site is a scheduled ancient monument. (HEW 0798)

183 *Middle section of the locks, Rogerstone.*

S14. Newport Transporter Bridge (ST 318862) is one of the few remaining bridges of its type in the world. Across the River Usk, it is about 1½ miles south of Newport town centre and was completed in 1906. It is perhaps better described as an aerial ferry rather than a bridge. The bridge is a slender attractive structure. The only other similar bridges in Britain are at Middlesbrough and Warrington.

The development of industry south of the town needed an additional river crossing downriver of the town bridge. The design was chosen in order to give adequate headroom for the shipping using the River Usk at the turn of the century. Several alternatives had been considered but the flat topography and the need for height to clear shipping in the river precluded a conventional bridge. A lifting bridge was considered but the Borough Engineer, R.H. Haynes, was aware of a transporter-type bridge in France and commissioned the French engineer Ferdinand Arnodin to design one for Newport. The bridge has a span from centre to centre of the towers of 645ft and a clear headway from high water level of 177ft. The builders were the Cleveland Bridge & Engineering Company of Darlington.

The towers are of open lattice steel construction, each founded on four cast-iron cylinders. There are 16 suspension cables, four inside and four outside each of the stiffening girders, which are 16ft deep and 26ft 3in from centre to centre. The bottom booms are built-up plate girders, each lower flange carrying rails on which the travelling frame runs.

The platform car, which is 33ft long by 40ft wide, is hung from the travelling frame by 30 suspension ropes. The total weight of frame and car is just over 50 tons and

184 *The Transporter Bridge.*

185 *George Street Bridge from downriver.*

they are moved by steel wire ropes wound on a drum worked by two electric motors, each of 35 brake horse power.

The bridge was closed to traffic in 1985 because of corrosion, but was reopened in 1995 following extensive restoration. It is listed Grade I. (HEW 0144)

S15. George Street Bridge, Newport (ST 319877), was the first concrete cable-stayed bridge in the UK. It has reinforced concrete towers and a steel box deck. It is 1,770ft long between abutments with a centre span of 500ft. Overall width is 66ft including two nine-foot footpaths. Approach viaducts are in prestressed and reinforced concrete 620ft and 650ft long of flat slab construction. The hollow towers are 170ft high above caisson level and the caissons are 30ft diameter founded at 45ft below Ordnance Datum in red marl rock. The deck is supported by three-inch-diameter locked coil ropes at 38ft 4in centres. As there is no longer a need for headroom for commercial shipping, the air-draft has been reduced and the clearance under the bridge is much less than the transporter bridge downstream.

Construction was by the Cleveland Bridge & Engineering Company with steelwork by Fairfield Shipbuilding & Engineering of Chepstow. Construction commenced in 1962 and was completed in 1964. The bridge is listed Grade II*. (HEW 0346)

186 *Crumlin Viaduct in 1964. (© RCAHMW)*

S16. Crumlin Viaduct (ST 213986) was a major innovative wrought-iron railway viaduct. Of all the many viaducts in South Wales, it is perhaps the most interesting and certainly the most dramatic. When built it was the third highest in the world. It is still worth describing although it was demolished in 1966, one of the many cross-valley viaducts in South Wales that had served their purpose. Traces of the piers can be found on site and the massive abutments can still be seen at high level on the valley sides and are listed Grade II.

Crumlin Viaduct had 10 spans, divided into seven and three by a rock knoll on the western side. The total length approached 1,700ft. Crossing the valley of the Ebbw river at a height of 200ft, the viaduct was built in 1856 by the Newport, Abergavenny & Hereford Railway for their line to the Taff Valley. The piers each consisted of 12 cast-iron columns of 12in diameter in three rows of four, with an additional raker at each end covering an area of about 42ft by 21ft. Each column was in eight pieces with wrought-iron cross-bracing to adjacent columns. There were four lines of Warren truss girders, each of 150ft span and 15ft deep, cross-braced by iron plates (1868), replaced by steel in 1930. This was the first large-scale multi-span use of this design in wrought iron.

The viaduct was designed and built by T.W. Kennard, the partner of James Warren who designed the truss configuration which bears his name and which was first used for a major bridge at Newark Dyke on the Great Northern Railway main line. The Crumlin columns were cast at Kennard's works, the wrought iron supplied by the Blaenavon works, and were fabricated on site and hoisted into position. (HEW 0072)

187 *The pump at Melingriffith.*

Central South Wales

S17. Melingriffith water pump (ST 143800). The Glamorganshire Canal, which was opened in 1798, drew water from a feeder which also supplied the Melingriffith Iron Works in the Whitchurch area of Cardiff. The Canal Company was obliged to take its waters from the tail race below the Melingriffith Works at a lower level than the canal, and because of this a pumping engine had to be installed to lift the water 12ft into a canal feeder. It was originally intended to be steam-powered but the fast-flowing tailrace was seen as a more economical option to power a waterwheel. The original pump was probably designed by Watkin George and installed when the canal opened. In 1807 a proposal for a re-design was prepared by John Rennie and William Jessop. It operated for 135 years until the closure of the canal in 1942. The pump was well constructed of American oak and cast iron and survived largely intact until 1974 when a full restoration was undertaken.

The undershot water-wheel is 18ft 6in in diameter by 12ft 6in wide and consists of three cast-iron wheels mounted on an axle, originally of oak, now of steel. There are six cruciform cast-iron spokes to each and the rims carry 30 paddles, 22in deep. The wheel drove pistons in two cylinders via timber rocking beams 22ft long. The cylinders had a bore of 2ft 8in and a five-foot stroke.

The pump is a scheduled ancient monument located at the west side of Ty Mawr Road in Whitchurch about a quarter of a mile south of the end of the section of the canal still in water. (HEW 0392)

188 *Hengoed Viaduct.*

S18. Hengoed, or Maes y Cwmmer, Viaduct (ST 154949) is a substantial masonry railway viaduct, which, like the wrought-iron viaduct at Crumlin, was one of the major structures on the Taff Vale branch of the Newport, Abergavenny & Hereford Railway and was built in 1857. It has 16 semi-circular arch spans of 40ft, one skewed, and a maximum height of 130ft. It is built in rough stone, on a curve, and crosses the A469 main road, the River Rhymney and a minor road. The railway closed in 1964.

The viaduct was refurbished in 2005 and is now included in the National Cycle Route 47, the Celtic Trail, across South Wales. The parapets have been raised by the addition of a stainless-steel post and wire fence for additional security and a major sculpture was placed at the eastern end symbolising the many collieries that existed in the area. (HEW 0803)

S19. Taff Vale Railway (SO 052057 to ST 190750) was Brunel's first Welsh railway. The River Taff which rises in the Brecon Beacons is joined by the River Cynon at Abercynon and the River Rhondda at Pontypridd before flowing on to Cardiff. The Taff Vale Railway initially followed the river from Merthyr Tydfil to Cardiff. Later branches were built into the Rhondda and Cynon Valleys.

At the turn of the 18th and 19th centuries the discovery of iron ore along the head of the valleys and coal in the valleys led to the establishment of ironworks, such as Cyfarthfa and Dowlais at Merthyr Tydfil, and to a rapid expansion in coal mining. These industries required means of transport for the raw materials and for the finished products. The Glamorganshire Canal from Merthyr to Cardiff was built between 1790 and 1798 to give communication with the sea, but in general the South Wales countryside did not really lend itself to the development of canals and so there was a great proliferation of horse tramroads, such as the Merthyr Tramroad from Penydarren to Abercynon, linking to the main canals in the valleys.

By 1830 canals and tramroads were proving inadequate, so in 1835 the Merthyr ironmasters engaged I.K. Brunel to construct the first major commercial railway in South Wales. An Act of June 1836 authorised a single 4ft 8½in gauge track, 24¼ miles long with six passing places, from Merthyr Tydfil at the head of the Taff Vale to Cardiff, where dockland development was beginning.

Major works included a rope-worked 1 in 19 incline 45 chains (about 3,000ft) long north of Navigation (Abercynon) (ST 085951 to 090960); a masonry viaduct at Newbridge (Pontypridd) (HEW 1220) (ST 070901) with a skew span of 110ft across the River Rhondda; a short tunnel south of Goitre Coed (ST 090962); and the six-span heavily skewed masonry viaduct across the River Taff and the Merthyr Tramroad at Goitre Coed, Quaker's Yard (**S20**). The line was opened in 1840-1. It was dualled in 1862 as the need for additional capacity grew. The incline was replaced at a 1 in 40 gradient and a 70ft-deep cutting in 1867 and the tunnel opened out into a cutting.

In the course of time, coal became the dominant traffic. The Taff Vale Railway expanded into the Cynon valley from 1845 and by 1856 it had reached Treherbert near the head of the Rhondda Fawr. Before the end of the century the company had penetrated to Llantrisant, Cowbridge and Aberthaw in the Vale of Glamorgan and beyond Cardiff to Penarth, where it owned the docks which gave the railway satisfactory outlets to the sea.

The Taff Vale, being first in the field, was able to choose the easiest routes in difficult country. Later railways competing for the lucrative coal traffic had to contend with harsh topography, sometimes having to move from valley to valley, and at the expense of long viaducts and tunnels. It is little wonder then that the surviving railway lines in the area largely use the pioneering routes of the Taff Vale Railway. (HEW 1219)

S20. Goitre Coed (Quakers Yard) Viaduct (ST 089965) is a masonry arch viaduct on the Taff Vale Railway and a fine example of Brunel's engineering design.It is a heavily skewed and curved six-span crossing of the River Taff. Brunel was concerned about

189 *(Far left) The piers are capped with a substantial pediment and the arches themselves have substantial chamfers which repeat the angled faces of the piers. (©Owen Gibbs collection)*

190 *(Above right) The Taff Trail looking south on the line of the Merthyr Tramroad at Goitre Coed.*

scour around the pier foundations. After careful consideration he adopted the novel idea of octagonal pier bases so that the pier is still at right angles to the flow.

The viaduct was built in 1841. It was later widened in 1862 by another viaduct of less architectural merit on the downstream side. It is listed Grade II.

It is just south of Quakers Yard station but it is best seen from the Taff Trail which at this point is on the line of the **Merthyr (Penydarren) Tramroad (S25)** which passes under the easternmost arch. (HEW 0801)

S21 William Edwards' Bridge, Pontypridd (ST 074904) was the longest span masonry bridge in the UK when built. No study of bridges – especially masonry arch bridges – would be complete without a reference to William Edwards' famous arch across the River Taff at Pontypridd (called Pont-y-ty-Pridd at that time). He was commissioned in 1746 but his first attempt, a multi-span bridge, was washed away in a flood in 1750 after

191 **A***n engraving by J.C. Canot of a painting by Richard Wilson, c.1767. (© ICE)*

only two years. Edwards then decided to span from bank to bank, but the temporary support for the arch was washed away in a flood in 1751 even before the centre was struck. His next attempt was also doomed to failure. He attempted a single span of 140ft with a rise of 35ft, which meant a great weight of filling over the haunches compared with the crown where there was only the arch ring and the parapets. During the construction of the spandrel walls the excessive weight near the abutments forced the crown upwards and the bridge again collapsed. Fortunately this was not a sudden failure and Edwards had time to observe the mode of collapse.

Jervoise has described the last and successful single span as follows: 'Edwards then rebuilt the bridge to the same design except that he placed at each end three cylindrical holes graduated in size, the largest being nine feet in diameter, to relieve the arch from the pressure of its haunches.' The spandrel infilling was of charcoal, for further lightness. This scheme proved successful and the bridge, which was completed in 1755, still stands. It remained the longest masonry arch bridge in the UK until the construction of Over Bridge in Gloucester in 1828. A commemorative plaque is inscribed 'William Edwards 1750. Repaired by David Edward and Thomas Evan 1798'.

The bridge soffit is an almost perfect arc of a circle of 89ft radius and the arch ring has a depth of construction of only 2ft 6in. It is 11ft wide between parapets. The relatively large rise at the crown resulted in steep slopes at either end of the bridge and this caused serious problems for heavy carts, both during the ascent and descent. These were restrained by a system of chains and a counterweight on the descent.

A modern bridge has been built alongside and Edwards' masterpiece is preserved for use by pedestrians only. The bridge is a scheduled ancient monument. (HEW 0027)

192 *View of bridge, c.1980. (© Owen Gibbs Collection)*

S22. Berw Road Bridge, Pontypridd (ST 077911) was the longest span in reinforced concrete in the UK when built. It crosses the River Taff about half a mile upstream from William Edwards' masonry arch bridge and is one of several early reinforced concrete bridges in South Wales designed by L.G. Mouchel to the system developed by the French engineer François Hennebique. Mouchel was Hennebique's agent in South Wales; he lived in South Wales for most of his life and was for a time the French consul.

The bridge has a central clear span of 116ft and side spans of 25ft. The width between parapets is 26ft. The main span has three parabolic arched ribs at 12ft centres, braced at intervals. The longitudinal beams supporting the deck are in turn borne by columns off the ribs over the outer thirds of the span, and the arch itself serves as direct support to the deck over the middle third. The side spans have their outer main beams arched to match the centre span. It is listed Grade II*. The bridge was completed around 1909 by Watkin Williams & Page. The deck was reconstructed in the 1970s. (HEW 0620)

S23. Pont-y-Gwaith Bridge (ST 080975) is a masonry bridge of 55ft span, 15ft 9in rise over the River Taff about a mile north of Quaker's Yard. It has several features in common with Pontypridd, including the use of thin stones to form the arch ring, the severe road gradient and the narrowing in plan from the abutments to mid-span, but it has no opening in the spandrels. Both span and width of the bridge are much smaller than those of Pontypridd. A plaque records it was built in 1811 to replace an earlier timber bridge associated with 16th-century ironworks on the west side.

Following damage and distortion by mining subsidence that affected the arch shape, the structure was repaired in 1989 with a lightweight concrete saddle and refurbished in 1993. It now forms part of the Taff Trail. Listed Grade II, it is restricted to pedestrians and cyclists. (HEW 0800)

193 *Pont-y-Gwaith.*

S24. Pont-y-Cafnau Bridge (SO 038071) was one of the earliest uses of cast iron for a tramway bridge, with timber joint construction features. This unique 'bridge of troughs' still spans the River Taff where it was built in 1793 to carry a tramroad and water supply into the Cyfarthfa ironworks in Merthyr Tydfil. The designer was the chief works engineer, Watkin George. In 1795 a second bridge, which no longer exists, was cast from the same patterns to carry an extension of the tramroad from the works to the Glamorganshire Canal.

The bridge, now used by pedestrians, spans 47ft. Two substantial A-frames, one on each side of the deck, have their feet embedded in the river walls, with the apex at mid-span. The frames are held together by mortise and tenon and dovetail joints (George was a former carpenter) and incorporate sockets which carry transverse members at mid-span and at the quarter points. These in turn support the deck structure, which is a closed rectangular box about two feet deep and 6ft 2in wide that carried the leat inside with the tramroad on top.

Pont-y-Cafnau undoubtedly had its influence on other better-known structures. In 1794 the Shropshire ironmaster, William Reynolds, sketched the bridge and in the following year Telford reported that the proposed design and method of construction at Longdon-on-Tern Aqueduct, itself a prototype for Pontcysyllte, had been referred to Reynolds.

194 *Pont-y-Cafnau.*

The bridge is listed Grade II*. In recent years the structure has been refurbished by the local authority and this has exposed an interesting arrangement whereby the rail chairs are cast integrally with the deck. (HEW 0656)

S25. Merthyr (Penydarren) Tramroad (SO 056070 to ST 085950) is famous as the location of the world's first journey by steam locomotive. Usually known as the Penydarren Tramroad, it was engineered by George Overton and opened in 1802. It has a unique place in railway history. On 21 February 1804 the first journey by a steam locomotive was made on it, using a steam engine built by Richard Trevithick. The trial was successful and the engine made several journeys in the following weeks. However, the rails proved too fragile for the weight of the locomotive so that although the experiment showed the possibilities of this new form of traction it also demonstrated that these could not be

195 *Victoria Bridge.*

196 *Greenfield Bridge.*

fully realised until a much better track or lighter locomotive became available.

The tramroad is nine and a half miles long at an average gradient of 1 in 45. The L-section cast-iron rails were three feet long, weighed 56lb each, and were laid to a gauge of 4ft 4in, over the outside of the flanges. Later these were replaced with plateways at the standard gauge of 4ft 8½in. The National Waterfront Museum at Swansea has a working replica of the locomotive. The rails upon which it runs are replicas of the Penydarren Tramroad rails but laid on concrete blocks instead of stone.

The Glamorganshire Canal was completed in 1798 between Merthyr Tydfil and the sea at Cardiff. The section above Abercynon had a set of 10 locks and also a water supply problem which limited its capacity. Richard Crawshay (whose Cyfarthfa ironworks were on the west bank of the Taff) had first call on its facilities, which led three other ironmasters, Guest, Homfray and the Earl of Plymouth, with works on the east bank in Merthyr (Dowlais, Penydarren and Plymouth respectively), to build a tramroad from Samuel Homfray's Penydarren works to the canal at Abercynon. The *Navigation Hotel* nearby was once the offices of the canal company. As the Act authorising the canal also authorised connecting tramroads no extra authority was needed. Most of the route is now a footpath and cycleway. Traces of the original stone pot sleepers can still be found.

197 *Goitre Coed Viaduct on the Taff Vale railway at the time it was being widened c.1862. The tramroad can be seen on the right with a horse-drawn tram waiting in a passing loop. (© John Minnis)*

The route northwards from its southern terminus at Abercynon (ST 085950) (at the modern fire-station building) runs initially along the eastern bank of the River Taff. Near Quaker's Yard the river was crossed twice by two timber bridges. In about 1815, following the collapse of one of them as a train passed over, these were replaced by segmental, nearly semi-circular, masonry arches of 60ft span and nine-foot width. These have parallel rings about 18in deep, composed of two- to three-inch flat stone voussoirs similar to **Pont-y-Gwaith (S23)**. (Penydarren Tramroad Bridges, HEW 0799/01 and /02 at ST 090966 and ST 094963).

Further north the tramroad was crossed in 1841 at Goitre Coed by the Taff Vale Railway on viaduct (**S20**) The section between Edwardsville, Quaker's Yard and Pont-y-Gwaith follows the deep valley of the Taff through pleasantly wooded country.

The southern portal of a tunnel as the tramroad enters Merthyr Tydfil at SO 058045 was exposed during land reclamation work in the 1980s and rebuilt with the addition of a mosaic wall mural and entrance floor. The tunnel was built probably as a cut-and-cover construction to allow waste material from the Plymouth ironworks to be deposited over it.

North of the tunnel the route continues on the east side of the road into Merthyr, where its line is perpetuated in the names of two streets, Tramroadside South and Tramroadside North. At SO 056070, near the point of the beginning of the tramroad, there is a memorial to Richard Trevithick. (HEW 0705)

S26. Cefn Coed y Cymmer Viaduct (SO 030076) is a 15-span curved masonry viaduct 115ft high. During the second half of the 19th century the network of railways expanded rapidly in South Wales to handle the vast tonnages of coal being produced. The difficult terrain forced the construction of numerous viaducts, some with masonry arches, some of iron. Often of great height and curved in plan, they produced some excellent examples of bridge engineering. Many have since been demolished and others are disused, such as Cefn Coed, a masonry viaduct which carried a part of the Brecon & Merthyr and London & North Western Joint Railways over the Taf Fawr from a junction north of Dowlais. Built on a curve, it is 770ft long and 115ft high with 15

198 *Track bed north of Edwardsville.*

199 *South portal of Trevithick's Tunnel.*

200 *Cefn Coed Viaduct.*

201 *Robertstown Bridge.*

semi-circular arches with spans of 39ft 9in in coursed rubble masonry with dressed quoins, the arches themselves being in brickwork. It was designed by Alexander Sutherland in consultation with Henry Conybeare and built by Savin & Ward in 1866. The curved alignment was to avoid crossing Crawshay land. The line opened on 1 August 1867 and closed in 1962.

The viaduct was refurbished in 1997 by the local authority, with grants from the National Lottery and Cadw. It is now part of the Taff Trail footpath and cycleway and is listed Grade II. (HEW 0171)

S27. Robertstown Bridge, Aberdare (SN 997037) is certainly one of the oldest surviving cast-iron railway bridges in the world. In 1811 the Aberdare Canal Company completed a tramway between ironworks at Hirwaun and the canal head at Cwmbach, Aberdare. This bridge carried the tramway across the River Cynon between Trecynon and Robertstown.

Four arched and trussed cast-iron beams spring from continuous cast-iron brackets built into the abutments. The width of each truss is only three inches and the depth varies from one foot at the centre to over five feet at the ends. Seventeen cast-iron plates 9ft 11in wide make up the total length of deck of 36ft 8in. The imprints of the rail fixings can be seen, showing it was single track. The stone abutments of bedded dressed stone with alternating courses of large and smaller stones were skilfully built. The bridge is now used as a footway and is listed Grade II. (HEW 0371)

S28. Barry Docks (ST 124668) is a complete major Victorian dock complex on a virgin site. In 1889 the Barry Railway Company built a dock between Barry Island and the mainland to capture a share of the export trade in South Wales coal which had developed enormously over the previous 50 years. The Engineer was John Wolfe Barry, one of whose assistants was I.K. Brunel's son Henry. Henry Brunel had produced earlier plans for a dock at this location. T.A. Walker was the contractor, following his success in completing the Severn Tunnel.

One of the principal supporters of the scheme was David Davies of Llandinam, already a well-known railway contractor, the owner of the Ocean Collieries and the leader of the Rhondda mine owners. His statue stands in front of the former dock office at Barry. Competition for space at the congested port of Cardiff and preferential treatment to the shareholders of that port stimulated the construction of alternative facilities by this group and Barry was selected as the site for a complete new docks complex. A railway line was also constructed to link to the Taff Vale Railway at Treforest in the Taff Valley and Hafod in the lower Rhondda valley. A second link was made to the TVR at Cogan just north of Penarth. In 1897 a line was built from Barry westwards to Bridgend to join the GWR.

The works over a period from 1884 to 1898 comprised a basin 500ft by 600ft and a dock 3,400ft by 1,100ft divided by a mole into two arms at its western end and covering 73 acres. The wrought-iron gates in the 80ft entrance lock to the basin and between the basin and the dock (which became No. 1 Dock) were the first to be operated directly by hydraulic rams instead of by opening and closing chains. At the north-east corner of the dock there was a dry dock 740ft by 100ft with a 60ft entrance closed by a caisson.

The dock was liberally supplied with hydraulic hoists for tipping coal wagons into ships. The excavation was generally left with sloping sides. Walls were provided only round the basin, along the south side of the dock and at the positions of the hoists; they are 46ft 6in high from dock bottom to coping, and were built in the dry, using massive limestone blocks. Elsewhere the dock sides are stone-pitched slopes.

Breakwaters were constructed off the entrance, mainly with rubble from the excavations, faced with four-ton stone blocks. On the end of the west breakwater is a cast-iron lighthouse 38ft high, 7ft 9in diameter at the base and 6ft 6in diameter at the top. This is a circular iron tower dating from 1890, 38ft high and made by Chance Brothers, which may be compared with **Whiteford Point Lighthouse (S27)**.

202 *Barry Docks Lighthouse and Breakwater.*

Before the works started, coffer-dams had to be formed by tipping across the channel to enclose the dock site and to enable the five million cubic yards of excavation to proceed. A Cornish engine of 267,000 gallons per hour capacity was brought from Walker's Severn Tunnel works to assist in the dewatering. A pier was built on the Bristol Channel side of the docks, with a passenger line from Barry Island station to the Pierhead station through a tunnel, both now closed. A second dry dock, similar to the first, was opened in 1893. The 34-acre No. 2 Dock was opened off No. 1 Dock in 1898, while in 1908 the Lady Windsor Lock, 647ft long and 65ft wide, named after the wife of the company chairman, was opened directly into the sea to the west of the basin.

203 *Glanrhyd Bridge and motorway viaduct. (© Owen Gibbs Collection)*

The coal trade reached its peak in 1911 when 11 million tons passed through the port but after a few years entered upon a continuous decline. Coal exports continued into the 1960s. The present owners, Associated British Ports, have adapted the docks to handle a variety of trades. The disused coal handling areas have been redeveloped for housing, retail and commercial use. (HEW 1233)

204 *(Left) Barry Docks in 1932. No. 1 dock in the foreground, No. 2 dock in the background and the basin and entrance lock on the middle right. (© RCAHMW –Aerofilms Collection)*

S29. Glanrhyd Bridge (SS 899828) is a 19th-century tramroad bridge. The Duffryn Llynfi tramroad linked ironworks and collieries in the vicinity of Maesteg to a small harbour at Porthcawl. A branch from Tondu to Bridgend was carried across the Ogmore river by means of a three-span masonry bridge known as Glanrhyd or Tyn y Garn Bridge. The arches span 23, 31 and 26ft and the width between parapets is 13ft.

The bridge, on which a stone plaque records that it was built in 1829, is well proportioned and of good workmanship. The designer is said to have been John Hodgkinson, engineer for the Dyffryn Llynfi Tramroad, and it was built by Morgan Thomas of Laleston. It now carries a minor road which turns off the Bridgend to Maesteg road, the A4063, to the north of Pen y Fai hospital. The bridge lies below a span of the Ogmore valley viaduct of the M4 motorway and served to support a section of the formwork during the construction of the viaduct. The two structures in such close proximity vividly illustrate the change in scale of transport provision over the last 150 years.

The tramroad was built to a gauge of 4ft 7in and was intended for horse-drawn traffic. In due course the lines were converted to standard gauge and from 1861 were worked by steam locomotives, although Glanrhyd Bridge itself was never used by these. Commemorative plaques now mark the termini of the tramroad at Maesteg, Bridgend and Porthcawl. The bridge is listed Grade II*. (HEW 1084)

S30. New Inn Bridge, Merthyr Mawr (SS 891784) is an ancient four-span masonry bridge, which takes its name from a hostelry that once stood nearby. It carried the main South Wales coaching route over the Ogmore river, now a minor road, B4265, just south of the A48 and one mile south-west of Bridgend. The bridge is of masonry, with four pointed arches. Each span is about 15ft and the width between the parapets is 10ft. The bridge cutwaters are triangular in plan, the upper parts stepped back in

205 *New Inn Bridge and sheep dip. (© TW)*

seven steps to terminate just below the top of the parapet, similar to other bridges in the area of the 18th and 19th centuries.

The parapets contain two openings through which sheep were once forced to leap into the deep pool beneath the bridge. For this reason the bridge, which is several hundred years old and is a listed Grade-II*scheduled ancient monument, is also known as the Sheep Dip Bridge. Jervoise suggests it may well have been the one seen by Leland in the 16th century. (HEW 1230)

Several other interesting bridges are nearby, including stepping stones, a suspension bridge and a reinforced concrete footbridge. Just downriver is another masonry arch bridge built by Morgan Thomas in 1827 to a design by William Whittington (SS 891779).

S31. Nant Hir Reservoir (SN 989068) is an 1870s earth-fill dam. The Afon Cynon, a tributary of the Taff, has several interesting reservoirs on it above Aberdare. One of these, built in 1875 for the Aberdare Waterworks, is on a side stream at Nant Hir. It has an earth dam some 63ft high and 328ft long on the crest, with a puddled clay core nine feet wide at the top, 21ft 6in at original ground level with an 8ft wide 17ft 3in deep cut-off trench, a grassed downstream 1 in 2 slope and stone pitched 1 in 3 upstream slope. It was the work of J.F. La Trobe Bateman, one of the most active water engineers of his day. He was the son-in-law of William Fairbairn, who developed tubular bridges in association with Robert Stephenson. The contractor was A. Sutherland, probably of Aberdare. Across the upper part of the Nant Hir reservoir is a notable reinforced concrete open-spandrel arch bridge carrying the A465, the Heads of the Valleys Road, built in the 1960s.

To visit the dam turn left (south) after the Baverstocks Junction on the A465, signed 'crematorium'. A lane 200m on the north after the crematorium leads to the dam and a footpath carries on to Nant Moel Dam in a further 1.2km. (HEW 1305)

S32. Upper Neuadd Dam (SO 030187) is the highest of the reservoirs on the Taf Fechan. There are six reservoirs on the Taf Fawr and Taf Fechan rivers, which flow from the Brecon Beacons and join to form the River Taff at Merthyr Tydfil. One of these is the Upper Neuadd, built in 1902 to the design of G.F. Deacon, who had previously completed the **Vyrnwy Dam (N34)**. The dam is of the gravity type, 77ft 5in high and 1,385ft long on the crest, and is unusual in that the entire dam is of a thin masonry section buttressed with earth fill over most of its length. It has an area of 59 acres and impounds 350 million gallons. The top water level is over 1,500ft above sea level.

The dam is founded on Old Red Sandstone strata, a factor which led to the choice of masonry rather than earth-fill construction. The central 89ft of the dam is flanked by towers and is curved to a radius of 310ft in plan to form a spillway. This section has a base width of 82ft as a result of the spillway. The rest narrows to 48ft. The exposed face is of massive limestone masonry blocks. The heart of the dam was built by bedding large masonry blocks laid on a thick mortar bed and voids filled with well-rammed mortar and stone rubble. The work was completed in 1902. (HEW 1241)

206 *Upper Neuadd Dam.*

S33. Usk Dam, Sennybridge (SN 833288), is an earth-fill dam of 1955, the first to have horizontal drainage blankets. Not far from the source of the River Usk, it rises some 109ft high and is 1,575ft long on the crest. It was built in 1955 to create a water supply reservoir for Swansea, by Richard Costain Ltd to a design by Binnie, Deacon & Gourlay. It is of historical interest because it was the first earth dam in Britain to be provided with horizontal drainage blankets in the embankment. This practice, suggested by Professor A.W. Skempton of Imperial College, London, has since become standard when clay fill is used. The earth-fill is boulder clay and the dam has a puddled clay core six feet thick at the top and 16ft thick at the base, below which is a concrete cut-off wall. The upstream face has a slope varying from 1 in 3 at the top to 1 in 4 at the base and is protected by concrete slabs.

The Usk flows eastward at this point and the water intake to carry water west to Swansea is on the river some distance from the dam site. The flow passes through a tunnel some 1½ miles long before emerging into a pipeline to cover the remaining 33 miles to Swansea. Since inception a water 'grid' has been established in South Wales and the reservoir now serves mainly to regulate the flow in the River Usk as a supply system to towns along its length. The reservoir is about five miles south of Trecastle and well signed from there. (HEW 1240)

207 *Usk Dam and spillway.*

South-West Wales

S34. Neath and Tennant Canals

S34.1 Neath Canal (SS 735941 to SN 090078) is the second major canal built in South Wales. It was constructed between 1791 and 1799 to bring iron ore down to Neath Abbey Works and for the export of coal. The earliest canal in the valley was Mackworth's Canal built between 1695 and 1700, really a tidal cut to bring vessels close to the Melyn Lead and Copper Works. The river Neath was navigable to Aberdulais where another artificial cut joined to the river by locks was built *c.*1700 to serve Ynys-y-Gerwyn tin and rolling mills. The Neath Canal itself was planned in 1790 for a canal from Pont Nedd Fechan at the head of the valley to Neath and wharves at Giant's Grave south of the town. Thomas Dadford junior, assisted by his father and his brother John, was commissioned and

208 *South end of Neath Canal at Giant's Grave.*

209 *Restored upper section of Neath Canal at Resolven.*

produced a scheme following the west side of the valley, using the river for part of the lower section from Ynysbwllog to Aberdulais and then continuing the canal on the east side to Neath. In 1791 this was modified to avoid the river and crossed to the east side at Ynysbwllog by aqueduct.

Thomas Dadford junior started work as General Surveyor on the lower section and Ynysbwllog was reached by mid-1792, when he left to build the Monmouthshire Canal. Thomas Sheasby was taken on as Engineer and contractor and continued to 1798, completing 10½ miles and 19 locks. The section from Neath to Giant's Grave was constructed by Edward Price of Govilon between 1797 and 1799. It was further extended south a few hundred yards in 1817, and a further half-mile in 1843.

There were three short branches, one towards Maesmarchog with a tramroad link to collieries, one towards Aberclwyd with a tramroad bringing limestone and coal from Cwmgwrach and one to Cwrt Sart south of Neath to collieries. The Aberdare ironworks was linked by tramroad to the head of the canal until 1812 when the Aberdare Canal was built accessing the Glamorganshire Canal instead.

The construction of the Vale of Neath Railway in 1851 on the eastern side of the valley affected trade but not as severely as anticipated and the canal continued to serve the western side of the valley. However, the canal slowly declined with little use after about 1883, closing eventually in 1934. One of the industries using the canal, a gunpowder factory at Glyn Neath, closed in 1931 which may have contributed to the final closure to traffic. The use of its water for industries, however, meant that much of the length remained in use for this purpose.

In recent times sections have been restored and further work is planned. At Resolven a section northwards has been fully restored and easily accessed from a nearby car park. The southern end at Giant's Grave has also been restored and is accessible for walkers from there to Neath. At Aberdulais both the Neath and Tennant canals can be accessed from a nearby parking area and the towpaths both north and south are readily accessible. (HEW 0570)

S34.2. Tennant Canal (SS 680935 to SS 774993) is the second longest privately owned canal and the most important built without Act of Parliament; it was formerly known as the Neath & Swansea Junction Canal and Red Jacket Canal. In 1817 George Tennant (1765-1832), son of John Tennant, a Lancashire solicitor, took an interest in the disused

210 *Lock at Aberdulais Aqueduct.*

Red Jacket Canal, built in the late 1780s by Edward Elton of Glanywern Colliery to link his colliery to the River Neath at Red Jacket Pill. It became disused after his bankruptcy and death in 1810. Tennant planned to clean and deepen it, extend it westward to Swansea and build a river lock at Red Jacket Pill. He leased the Glanywern Canal from the Earl of Jersey and with William Kirkhouse as Engineer work started in 1817, finishing in 1818 the four miles from the Neath to the Tawe. Work was carried out without a Parliamentary Act, landowners' consent being negotiated.

He changed his original plans for a lock at Giant's Grave to access the Neath Canal and instead began a length of almost five miles northwards from Red Jacket Pill, past Neath Abbey Works and through a 30ft-deep cutting to join the Neath Canal at Aberdulais. Again Kirkhouse was Engineer and in 1823 he started on the only lock on the 8½-mile main canal and on the 310ft-long Aberdulais aqueduct to reach the Neath Canal. This was completed in 1824. The river lock at Red Jacket Pill was little used after this.

Several short branches were constructed between 1828 and 1840 and in 1863 and 1869 to serve various iron and copper works and collieries. Port Tennant Tidal Dock at the western end of the Tawe at Swansea was converted into a wet dock and opened in 1881 as the Prince of Wales Dock.

211 *Aberdulais Aqueduct.*

212 *Pont Gam at the junction of the Neath and Tennant Canals, Neath Canal on the left, Tennant Canal on the right.*

The main traffic was coal, culm, some timber, iron ore, sand and a regular packet service. The canal closed in 1934 but was retained as a water supply to industry alongside it. The structures at Aberdulais – tollhouse, lock, aqueduct, basin and Pont Gam – are listed Grade II (SS 772993 and SS 774994).

Pont Gam is particularly interesting. It crosses the junction of the Tennant Canal with the Neath Canal and is 'Y' shaped in plan to connect the towpath from the north side of the Tennant Canal to the west side of the Neath Canal, allowing the Neath canal towpath to continue in both directions with the Tennant towpath also continuing northwards under the bridge. This area and the towpaths north and south are accessible from a nearby parking area.

S35. Swansea & Mumbles Railway (SS 657927 to SS 630875). The Oystermouth Railway, as it was originally known, was authorised by Parliament in June 1804 and was the world's first fare-paying passenger railway, opening in 1806. The rails were 'L'-shaped tramplates on stone blocks. The Act mentions haulage by 'men, horses or otherwise' – the 'otherwise' perhaps being with steam locomotives. The six-mile line was originally planned to link with the Swansea Canal and to carry goods traffic between it and Oystermouth, bringing limestone from Oystermouth and some coal from Clyne Valley. Users carried their own cargoes and paid a toll.

Edward Martin of Cumbria, a prolific civil and mechanical engineer and owner of Gwaunclawdd Colliery, made the survey and estimates and oversaw the construction. He was much involved in other enterprises in the Swansea Valley as well. He was mining agent for the Duke of Beaufort; he surveyed the line for the Swansea Canal and others in the area, as well as being engineer for the North Dock in Swansea.

The first load of passengers was carried on 25 March 1807, using horse traction. The passenger service ended about 1826 when a turnpike road opened from Swansea to Mumbles but was re-instated by a new owner in 1860, who reconstructed the line with standard edge rails.

Steam traction was introduced in 1877 and until 1896 was used in conjunction with horse power. Conventional side- and saddle-tank locomotives were used from 1878. In 1879 the line was renamed the Swansea & Mumbles Railway and in 1898 was extended to Mumbles Pier, which opened in 1898. It was electrified in 1929, being operated by double-decker tramcar type vehicles. Eleven Brush Electrical double-bogie two-deck tramcars were brought into service in 1929 and two more in 1930, each accommodating over 100 passengers, easily the largest built for passenger service in the UK. Service continued until 5 January 1960 when the line was closed on the basis that modern buses provided a more flexible service. In the 1950s the line was carrying more than three million passengers each year. A replica of the original horse-carriage and cab of the last Mumbles Railway tram can be found in the Swansea Museum Tramway Centre.

The track ran along the seaward side of the present A4067, which links the Gower Peninsula to Swansea. Although the tracks have now been removed, much of the line west of Blackpill can still be followed as a cycleway and footpath along the foreshore. The former Blackpill station, built in 1927, which also housed an electricity sub-station, can still be seen and remains in good condition as a café and restaurant. In March 1981 a stained-glass window was unveiled in Oystermouth parish church to mark the 175th anniversary of the opening of the railway. (HEW 0706)

S36. Mumbles Pier (SS 632874) was opened in 1898, designed by W. Sutcliffe Marsh. The contractors were Mayoh & Haley, with cast iron from Widnes Foundry Company. The Llanelly Railway obtained permission in 1865 to build a Mumbles Branch and

213 *Mumbles Railway and Mumbles Pier. (© Owen Gibbs Collection)*

214 *Mumbles Pier.*

Pier but it was never completed. Sir John Jenkins of the Rhondda & Swansea Bay Railway promoted a Bill enacted in 1889. Construction began in 1897.

The pier is 835ft long built on a substructure of cast-iron piles below a timber deck carried on steel lattice girders. On each side are three refuges, two of which carried pavilions, and at the seaward end it broadens out and at one time had a bandstand. There is a lower stage for embarkation to pleasure steamers which was rebuilt in 1956. A lifeboat station was added to the north side, accessed via a lattice girder walkway, also in 1956. (HEW 2693)

S37. Whiteford Point Lighthouse (SS 443973) is the only cast-iron wave-swept lighthouse in the country. A scheduled ancient monument listed Grade II*, it stands offshore between high- and low-tide levels on the north-west tip of the Gower peninsula and looks across the Burry Inlet to Burry Port and Llanelli, whose redundant harbours have now been brought back into use as leisure facilities. It was built in 1865 under the Llanelly Harbour Act of 1864 to replace an earlier light, carried on timber piles, which had lasted only a few years. The light was extinguished in 1921.

The lighthouse stands just above the low water mark and is 44ft high. Eight courses of cast-iron plates, bolted together through internal and external flanges, form the circular tower which has the curved taper typical of many lighthouses. The circular base is 24ft diameter and as the diameter decreases upwards to about 11ft 6in at lantern level care has been taken to reduce the width of the plates to stagger the vertical joints. From the seventh level 10 sturdy cast-iron brackets support the main balcony, which is of iron balusters linked at the top with trefoils and carried on 6ft 6in iron beams. The lantern is surrounded by an ornate wrought-iron balcony. At some time after erection the tower was strengthened by wrought-iron bands. Hennett & Spink of Bridgwater supplied the ironwork. They also supplied the ironwork for the much smaller one on Porthcawl breakwater.

The area to the south of the lighthouse, which is some two miles from the nearest road, is in the care of the National Trust. The lighthouse is accessible on foot at low tide. (HEW 1256)

215 *Whiteford Point. (© TW)*

S38. Llanelli Railways is a group of early railways replacing canals and tramways. The three railways described here were all based on earlier transport systems, the first being on the route of a canal and the other two following the lines of old horse tramways. This area of Carmarthenshire is rich in anthracite coal deposits which have been mined and exported for many years.

S38.1. Burry Port & Gwendraeth Valley Railway (SN 451008 to SN 529125). In 1766, Thomas Kymer obtained powers for a three-mile canal from Kidwelly Harbour. The first Welsh canal to obtain an Act of Parliament, this canal eventually extended north-eastward up the valley of the Gwendraeth Fawr as far as Cwm-Mawr and south-eastward to Burry Port. At Burry Port a tramroad led east to a junction with the Carmarthenshire

Railway near Llanelli. In 1865 the Kidwelly & Llanelly Canal Company obtained powers to abandon the canal and construct a railway on its line, and in 1866 it amalgamated with the Burry Port & Gwendraeth Valley Railway Company, which had been established in the same year. Sir William Shelford and John Robinson were the Engineers for the conversion.

The railway line followed the route of the canal and its tramway from Kidwelly Harbour, passing under the Great Western Railway's South Wales Railway by a low bridge before running northward to a terminal station at Cwm-mawr. From a triangular junction at SN 428059, the route to Burry Port and Llanelli ran south-eastward, crossing the GWR line again at Burry Port. Low bridges were a characteristic of the line, betraying its canal ancestry, for many of the original canal bridges were not modified for the railway and specially cut-down locomotives have always been necessary to work the traffic. Passengers were carried from 1909 when a Light Railway Order was obtained. The line was absorbed by the Great Western Railway in July 1922.

The canal had been constructed in the period 1815-39. James Green was the Engineer and the canal included three inclined planes at Pont Henri (SN 476092), Capel Ifan (SN 491104) and Cwm-mawr (SN 523121). The railway made use of two old canal aqueducts, one over the Gwendraeth Fawr on the Llanelli line, **Gwendraeth Fawr Bridge (S39)** (HEW 1234, SN 428054). The second aqueduct is at SN 447073, near Pont Newydd. Of the former inclines, the first, Pont Henri, was replaced by a 1 in 48 grade on the same alignment. A diversion was needed at Capel Ifan and at Cwm-mawr a steep grade of 1 in 32 was used.

Passenger services on the line were withdrawn in 1953 and general freight traffic ceased in 1965. The GWR connection at Burry Port to Cwm-mawr survived, bringing coal from the Cwm-mawr opencast disposal point. This traffic ceased in March 1996 and the line is closed completely. (HEW 1217)

S38.2. Llanelly & Mynydd Mawr Railway (SS 500995 to SN 563131). The second line, the Llanelly & Mynydd Mawr Railway (L&MMR), had its origins in the Carmarthenshire Railway or Tramroad, a horse-drawn line running from the harbour at Llanelli northward to limestone quarries near Castell-y-Graig (SN 603164). It was only the second public railway to be authorised by an Act of Parliament, on 3 June 1802, the first having been the Surrey Iron Railway of 1801 (HEW 1387), and was possibly the first to come into service. It had flanged rails on stone blocks. Its designer was James Barnes, a well-known English canal engineer. Its length of 16 miles was quite long for a railway at that time. Never commercially successful, it closed in 1844.

In 1877 the L&MMR was incorporated to lay a standard-gauge railway on the old trackbed. It opened for freight traffic on 1 January 1883 and by 1887 was extended 13 miles to Cross Hands. There was no public passenger service but large numbers of colliery workers were carried in workmen's carriages. Originally the line crossed over the GWR at Llanelli to reach the docks, with a connection to the main line railway at the crossing point, but the docks line was later taken up. In 1923 the L&MMR was absorbed by the GWR. After 1966 the line was not used north of Cynheidre Colliery (SN 495075) and with the closure of this colliery the whole line is now disused. The track has been lifted except for a few sidings at the closed Cynheidre Colliery, where some rolling stock and a large shed are held by the L&MMR Railway Society. At the south end through Llanelli the track has been abandoned from the junction with the main line as far as SN 499005. From there to Cross Hands the track bed is in use as National Cycle Route 47 and most earthworks and structures can still be seen. North of Cross Hands little trace remains. (HEW 1388)

S38.3. Llanelly Railway (SS 532989 to SN 632217). The third railway, the Llanelly Railway, is still partly in use as a section of the Central Wales line from Craven Arms in Shropshire to Llanelli and Swansea. Its origins lie in the Llanelly Railroad & Dock Company, which was incorporated in 1828 to make a wet dock at Llanelli and a horse tramroad. The company became the Llanelly Railway & Dock Company in 1835. By 1839 under George Bush it had built a narrow-gauge railway from Llanelli to Pontardulais and this was changed to standard gauge and extended in stages to reach Llandeilo in 1857. Here the company leased the Vale of Towy Railway and so reached Llandovery in April 1858. Branches were opened into the Amman valley from Pantyffynnon in the 1840s and to Carmarthen in 1865, and in 1867 a further line from Pontardulais to Swansea opened, crossing the Gower and running along the shore eastward to Swansea (Victoria) station.

In 1868 the LNWR Central Wales line reached Llandovery from the north and the LNWR was granted running powers over the Llanelly Railway which enabled it to run into Swansea from January 1873. The Llandeilo to Carmarthen and Pontardulais to Swansea lines were absorbed by the LNWR in 1891 and the remainder of the Llanelly Railway passed into the hands of the GWR in 1889.

Only the original main line from Llandovery to Llanelli remains open, carrying a service of trains from Shrewsbury that reach Swansea along the GWR main line by a reversal of direction at Llanelli. A 17-mile-long mineral branch from Pantyffynnon is still in place as far as Gwaun-cae-Gurwen, where an opencast mine is still being worked.

There is an interesting short tunnel at Pontardulais (SN 587039), said to date from the horse-drawn tramway era, which had the invert lowered to accommodate standard-gauge traffic. (HEW 1389)

S39. Gwendraeth Fawr Bridge (SN 428054) is a canal aqueduct converted to rail use. Known also as Glastony Aqueduct by Cadw and also Trimsaran Bridge this aqueduct was constructed by the Kidwelly & Llanelly Canal Company in 1815 (HEW 1217). A plaque, now missing, recorded 'built under the direction of Messrs Pinkerton and Allen in 1815'. In 1837 the canal was incorporated in a more extensive network extending up the Gwendraeth Valley. In 1865 when the canal was converted to a standard-gauge railway, the existing iron trough was removed and the void filled with copper slag and masonry. The conversion was probably constructed by James Green.

It has six spans, each of 11ft, being segmental arches with the crown only two feet above the springing. The spandrel walls extend to about 11ft above springing level and the rail level is a further two feet above that. The two faces differ in detail. On the upstream face the arch voussoirs have their upper edges finished to a horizontal line, which runs across the face intersecting the tops of the cutwaters;on the downstream face the piers project from the spandrel face and have horizontal caps. It is understood that sluices were originally attached to the face of the bridge to prevent tides penetrating further upstream and these flat caps and rebated spandrel were the supports. The bridge is 40ft 6in wide but there seems no apparent reason for such a wide structure. The bridge is a scheduled ancient monument.

216 *Aqueduct c.1980 with track still in place. (© Owen Gibbs Collection)*

The bridge can be accessed via the old track bed. An unclassified road leads west from the

217 *Pont Spwdwr. (© Society for the Protection of Ancient Buildings)*

B4317 Pembrey to Llandyry road. Just after the *Plough Inn* the line crossed the road at SN 420020 and it is possible to walk from there a mile or so north along the track bed and alongside the disused and overgrown canal which is on the east side. (HEW 1234)

S40. Pont Spwdwr, Llandyry (SN 433058) is a medieval arch bridge, rare in the area. It has three arches with spans of 21ft, 28ft and 31ft together with a further three flood-relief arches of spans 19ft, 18ft and 10ft. The arches are of pointed segmental form and four intervening cutwaters are carried upwards to form pedestrian refuges at road level. The width between parapets varies from 12ft on the approaches to 8ft 5in on the bridge. It is constructed of roughly squared masonry and was built probably in the second half of the 16th century. The wing walls were added in 1770 for Thomas Kymer. The road was diverted in 1966 over a new bridge just upstream and the original bridge, which is a scheduled ancient monument, has been retained for pedestrians. (HEW 1218)

S41. Twrch Aqueduct, Ystradgynlais (SN 773092) was the first recorded use of hydraulic lime concrete to waterproof an aqueduct. When the Swansea Canal was being built towards its final termination at Abercraf, it had to cross the Afon Twrch at Ystalyfera, a tributary of the River Tawe. The Engineer, Thomas Sheasby, designed a three-span masonry aqueduct with a channel 11ft wide and four feet deep. To waterproof the channel he used hydraulic lime concrete instead of puddled clay, reputedly the first use of this material for this purpose. The three segmental arches have spans of about 30ft and the two piers have cutwaters on both sides. The cutwaters on the downstream side have a conventional triangular shape and originally the upstream cutwaters were of similar construction. However, because of the damage caused by boulders brought down the Afon Twrch in times of flood, the upstream cutwaters were later rebuilt into the substantial curved structures seen today.

218 *Twrch Aqueduct and weir.*

Immediately below the aqueduct the river falls over a stone weir about 13ft high. This weir served not only to protect the foundations of the piers from scour but also diverted water into a feeder channel to the canal on the south-west side.

The aqueduct had a central spillway on the upstream side and stop plank grooves at both ends. A sluice on the upstream side of the channel enabled the aqueduct to be drained for maintenance. On the north-east side of the aqueduct is a culvert which carried water from the tail race of a nearby fulling mill.

The Twrch Aqueduct was completed in 1798. The canal was sold to the Great Western Railway in 1873. Trade ceased on the section including the aqueduct before the end of the 19th century but continued on the lower canal until 1931. Although this section of the Swansea Canal has now largely vanished, the aqueduct has been preserved. It has been refurbished to serve as part of the National Cycle Route 43. Little is visible of the canal each side of the aqueduct, much having been buried under road improvements. (HEW 2074)

S42. Llandeilo Bridge (SN 627220) is the largest masonry span in Wales, with a clear span of 144ft 9in and an overall width of 33ft. It was built in 1848 and carries the trunk road A483 on a falling grade from the town over the Afon Tywi and its neighbouring water meadows. The single masonry arch has a rise of 35ft. It exceeds the span of William Edwards's bridge at Pontypridd by 4ft 9in. An earlier bridge on the site was a seven-span structure of some antiquity, possibly 16th-century, which was partly destroyed by floods by the late 18th century. The damaged three centre spans were replaced with a wooden bridge, itself washed away in 1798 and replaced.

219 *'Llandeilo Bridge' by M.A. Rooker, 1797, showing the damaged centre spans and the timber replacement. (© National Library of Wales)*

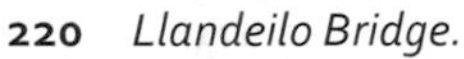

220 *Llandeilo Bridge.*

221 *The tide mill from downriver.*

The voussoirs are narrow but elongated and extend some way up the spandrel face. They are dressed to an ashlar finish with chamfered edges. The spandrel faces of the bridge, the arch soffit and the parapets, pilasters and wing walls are of squared coursed masonry, with the faces chiselled or hammer-dressed to an approximately flat surface. Pilasters occur at each end of the bridge and at intervals along the approach walls.

The bridge was designed by William Williams, the County Bridge Surveyor, and Edward Haycock, County Surveyor of Shropshire. Construction was started in 1844 by Morgan Morgan of Cwmamman, who failed to complete, and in 1846 the county direct labour took over and completed in 1848.

Perhaps in style it is too formal for such a rural setting but it has been described as probably the finest one-arch bridge in Wales. There are interesting references in the diary of Thomas Jenkins of Llandeilo to its construction, on which the diarist worked. (HEW 0681)

S43. Carew Tide Mill (SM 041038) is one of the few remaining tide mills in working order. A large three-storey stone building, it stands at the southern end of a dam across the upper estuary of the Carew River. Records of the mill can be traced back to Elizabethan times.

The dam, which has stone facings and a central clay core, impounds an area of water of some 27 acres replenished at high tide and this water is discharged through passages in the dam beneath the mill to operate undershot water-wheels of wood and iron construction. Each wheel is considered to have been able to generate some 20-brake horse power.

After a long period of disuse the mill was saved from imminent collapse in the early 1970s, largely through the efforts of Mr John Russell FSVA, and with financial assistance from various organisations. Following restoration by the National Park Authority the mill now houses an interesting milling museum. In the 1930s the sites of some 25 tide mills were recorded in southern Britain. Today few are left and even fewer are still in working order. (HEW 0803)

GLOSSARY

abutment – the masonry at each end of the arch which provides the resistance to the thrust of the arch.

accumulator, hydraulic – a large vertical cylinder/piston to pressurise a water power system; a heavy weight is raised to the top of the piston and the system is pressurised by the weight operating under gravity.

agger – a raised causeway upon which a Roman road was constructed.

aqueduct – a bridge carrying a watercourse over another watercourse, road, railway or valley.

arch dam – of masonry or concrete, used where the rock formations at the sides of the valley are able to support the thrust from the arch form, providing a more economical material and capable of being built to greater heights. Sound rock is also needed for the foundations.

arch types:

elliptical – similar to segmental but with the centre portion flattened;

pointed – two half-segments with radii displaced to opposite sides of arch, meeting at the centre at a sharp point, can be sharp or flat;

relieving – an arch built into a structure to reduce the load on a wall below;

semi-circular – formed of one half of a circle, rise equal to half the span;

segmental – formed of less than half a circle, rise less than half the span;

Tudor – each side formed of a short vertical length leading to a sharp curved corner and an upward-angled straight length to the centre.

asphalt – a high-strength road material of stone and binder of bitumen or natural asphalt.

balanced cantilever construction – the bridge is built in increments each side of a support pier. Construction can be either by casting the increments *in situ* using specially designed shuttering to form the hollow concrete sections or by casting the sections off-site and transporting and lifting into place. The method is particularly useful where access to the land under the bridge is difficult. Also used for steel bridges.

bascule – a single- or double-leaf bridge usually over a waterway which can be pivoted about a horizontal axis to give a clear headroom for vessels to pass through.

bitumen – a by-product of distillation of crude petroleum oil.

box girder – a girder of rectangular or trapezoidal cross-section whose web and flanges are relatively thin compared with the space in the box.

bowstring girder – a girder with a straight bottom chord from which springs an arched top chord, the two connected by rigid stiffening members.

breast shot wheel – a water wheel which is supplied by water at about mid-height.

buck – the body of a post mill.

buttress dam – a development of the gravity dam with the downstream face supported by buttresses, reducing the amount of material necessary to resist the water pressure.

bye-wash – a channel which allows excess water to by-pass a lock.

cable-stayed bridge – a bridge whose deck is directly supported by a fan of cables from the towers.

cantilever – a beam not supported at its outer end.

cast iron – an iron-carbon alloy with impurities which preclude it being rolled or forged; it has to be poured in molten form into moulds of the required shape and size.

centering – a temporary structure to support an arch during building.

conduit – an artificial watercourse, either in pipe or open channel, to convey drinking water. Often used also for the fountain, pump or other outlet from which the water is drawn.

crowntree – a great transverse timber beam at the top of the post in a post mill which rotates the buck and the sails to face the wind.

cut and cover – a tunnel or part of a tunnel constructed by excavating a cutting to build it and then burying it; often used at the ends of a tunnel to reduce the amount of actual tunnelling required.

cutwater – a v-shaped upstream face of a bridge pier; designed to part the flow of water and prevent debris accumulating around the pier; sometimes provided also on the downstream face.

dentilation – ornamentation resembling teeth, used to decorate cornices.

dripmould – a projecting course of stone or brick designed to throw off rain.

edge rail – a rail without an upstanding flange as opposed to a plate rail (q.v.).

embankment dam – resists the water pressure by its weight. Can be of compacted earth or rockfill with a vertical impermeable central clay core to prevent seepage through the dam. This form is particularly suitable for wide flat valleys. Rockfill types can have an impermeable upstream face rather than a core.

Engineer – the engineer in charge of the project.

engineering brick – brick with greater crushing strength than common brick.

extrados – the exterior curve of an arch, measured on the top of the voussoirs.

falsework – a temporary structure to support the formwork (q.v.) for casting concrete.

formwork – (or shuttering), a temporary structure to contain wet concrete and form its finished shape until set and self-supporting.

gauge – the distance between the inner faces of the rails for edge rail track or the outer flanges for a plateway:
standard gauge – the track gauge of 4ft 8½in in general use in Britain;
broad gauge – a track gauge greater than standard, up to 7ft ¼in, used by Brunel;
narrow gauge – any gauge less than standard, usually around two feet.

girder – a beam formed by connecting a top and bottom flange with a solid vertical web.

gravity dam – constructed of masonry or concrete resisting the water pressure by its weight, but because of the greater mass of the material the downstream and upstream slopes can be quite steep. This form is suitable for narrower valleys with sound rock for a foundation.

groyne – a timber, steel or concrete wall built at right angles to the shore to prevent or reduce littoral drift.

historical periods:
early medieval – A.D. 410 to 1066;
medieval – 1066 to 1536
post-medieval – 1537 to 1900
modern – 1900 to present day

immersed tube – prefabricated units in steel or concrete laid in an excavated trench in a river bed, jointed and covered over to form a continuous tunnel.

impost – the top part of a pillar, column or wall which may be decorated or moulded and on which a vault or arch rests.

incrementally launched construction – a bridge is cast in short lengths on the bank and a section pushed out over the valley using hydraulic jacks. As one section is pushed out another is cast behind and stressed to the previous one before it itself is jacked outwards. Permanent and temporary support piers are constructed in advance in the valley. The process continues until the bridge deck is completed across the valley. The bridge is then finally stressed to its design load. Also used for steel bridges.

intrados – the inner curve of an arch, also known as the soffit.

keystone – the central voussoir, often decorated.

leading lights – two lights by which a vessel may be aligned for safe entry into a harbour, usually one inland at high level and one lower down.

lintel – a horizontal beam over an opening.

littoral drift – the movement of sand or shingle along a coastline by the action of tides striking the coast at an angle and scouring the beach material.

loads: dead load – the total weight of the bridge structure itself;
live load – the weight of traffic crossing the bridge and loads from wind, snow etc.

locks: flash lock – a single gate in a river weir which can be opened to allow boats to pass through;

pound lock – an enclosed chamber with gates at both ends for moving boats from a higher to a lower level and vice versa.

mass concrete – concrete without added steel reinforcement.

navigator or navvy – labourer employed to excavate a canal, later any general labourer.

oculus, oculi – a round window or opening.

O.D. – Ordnance Datum, the mean sea level at Newlyn, Cornwall, from which Ordnance Survey measures heights in Britain.

order – where bridges have two or more arch rings and one stands forward of the one below, the arch rings are said to be in one (or more) orders.

overshot wheel – a water wheel which is supplied by water at the highest point of the wheel and turns in the direction of the flow (see pitchback).

pavement – the main traffic-carrying structure of the road
flexible – a pavement constructed of asphalt or tarmacadam, or stone or concrete blocks;
rigid – a pavement of mass or reinforced concrete.

pediment – the triangular termination of the end of a building etc. over a portico. Similar to a gable but with a less acute angle at the top.

penstock – a valve controlling a flow of water in a large diameter pipe.

pierre perdue – quantities of stone tipped loose into water to find their own position and form a breakwater.

pitchback wheel – a water wheel fed at the top but turning in the opposite direction to the flow, more efficient than an overshot wheel (q.v.).

plate rail – a rail, usually 'L'-shaped, on which travelled flangeless-wheeled waggons of tramways and early railways, the vertical flange of the rail providing lateral guidance.

portal frame – a frame of two vertical members connected at the top by a horizontal member or two inclined members with rigid connections between them.

post mill – a windmill in which the whole of the upper part including the sails and machinery is housed in a timber building all of which rotates on a large vertical post to face the wind.

post-tensioned concrete – prestressed concrete in which the reinforcement is contained in ducts cast in the concrete and tensioned after the concrete has set by jacking the bars or cables against anchorages cast at each end of the concrete unit.

prestressed concrete – concrete containing steel bar or wire reinforcement that has been tensioned before the live load is imposed. By utilising the concrete more effectively than in conventional reinforced concrete, lighter member sections can be used for a given load, though the quality of materials and workmanship required is higher

reinforced concrete – concrete containing steel reinforcement, thus increasing its load-bearing capacity

rolling lift (Scherzer) – a type of bascule developed by William Scherzer in America in 1893. As well as lifting, it also rolls back at the pivot point. It had advantages over the standard bascule bridge as it allowed the bridge to span greater distances, providing greater clearance over the waterway when it was rolled back.

roll-on/roll-off ferry – A ship which is loaded and discharged by vehicles driving on and off via a ramp (abbrev. ro/ro).

side pond – a small reservoir at the side of a pound lock into which some of the lock water can be discharged or recovered to reduce water use at the lock.

side pound – a widened area on a canal, usually situated where the distance between successive locks is very short, in order to provide extra water storage space.

sluice – in dams a system to release water at times when the reservoir level is below the spillway so as to maintain flows in the river or stream below the dam site; in water courses a barrier that can be raised or lowered to control the flow in the channel.

smock mill – a windmill having a timber tower resting on a stone or brick base. The top of the tower (cap) containing the sails rotates.

spandrel – the triangular area between the arch and the deck of an arch bridge.

spillways – release water when the reservoirs are full.

squinch – a small arch formed across an angle of a building or bridge.

stanch, staunch – a flood gate or watertight barrier.

statute labour – the requirement of the 1555 Act for parishioners to work for a fixed number of days each year on road maintenance.

stench stack – a hollow column from the top of a sewer to some height above ground level to allow ventilation of the sewer.

stringcourse – a projecting course of masonry or brickwork, often framing an arch or below the parapet of a bridge.

summit canal – a canal which passes over high ground and falls in both directions from the summit.

suspension – in suspension bridges the deck is supported by hangers from a suspension cable supported by towers and anchored to a foundation behind them.

tarmacadam or tarmac – a mixture of stone and a binding and coating agent of tar for road surfacing, or of bitumen (bitmac).

tower mill – a windmill with a brick or stone tower. The top of the tower containing the sails rotates, the machinery in the tower remains stationary.

truss – a girder in which the web is formed of discrete vertical and/or inclined members. There are many different configurations.

tub boat – an unpowered canal boat of less than normal dimensions carrying a few tons; often towed in trains.

undershot wheel – a water wheel supplied with water at the bottom of the wheel.

valve tower – built into a dam or nearby for the stored water to be taken off into the distribution pipe network or aqueduct for transmission to the treatment plants.

voussoir – the individual shaped blocks forming the arch.

wallower – the main gear in a wind or water mill which converts the rotation in a vertical plane of the sail or wheel into rotation in a horizontal plane to drive the machinery.

windpump – a windmill where the machinery drives a scoop wheel or turbine to raise water.

wrought iron – an iron-carbon alloy with few impurities capable of being rolled or forged as plate or bar.

BIBLIOGRAPHY

NB: In addition to the books used, information may be found for historical engineering works built after the mid-18th century in the records of the Proceedings of the Institution of Civil Engineers and of other professional bodies.

Abbott, W., *The Turnpike Road System in England and Wales, 1663-1840*, Cambridge University Press (1972)

Adamson, S.H., *Seaside Piers*, Batsford, London (1977)

Addis, W.A., *Building: 3000 Years of Design, Engineering and Conservation*, Phaidon, London (2007)

Baldwin, P. *et al.*, *The Motorway Achievement*, 3 vols, TTL, London (2002-9)

Balkwill, R. and Marshall, J., *The Guinness Book of Railway Facts and Feats*, Guinness Publishing, London (1993)

Barrie, D.S.M., *A Regional History of the Railways of Great Britain, Volume 12: South Wales*, David and Charles, Newton Abbot (1980)

Baughan, P.E., *A Regional History of the Railways of Great Britain, Volume 11: North and Mid Wales. Volume 12: South Wales*, David and Charles, Newton Abbot (1980)

The Chester & Holyhead Railway, David and Charles, Newton Abbot (1972)

Beaver, P., *A History of Lighthouses*, Peter Davies, London (1971)

A History of Tunnels, Peter Davies, London (1972)

Beckett, D., *Brunel's Britain*, David and Charles, Newton Abbot (1980)

Stephenson's Britain, David and Charles, Newton Abbot (1984)

Berridge, P.S.A., *The Girder Bridge*, Maxwell, London (1969)

Biddle, G., *Victorian Stations*, David and Charles, Newton Abbot (1973)

The Railway Surveyors, Ian Allan, London (1990)

Binnie, G.M., *Early Dam Builders in Britain*, Thomas Telford, London (1987)

Early Victorian Water Engineers, Thomas Telford, London (1981)

Blower, A., *British Railway Tunnels*, Ian Allan, London (1964)

Bowen, R.E., The *Burry Port and Gwendreath Valley Railway and its Antecedent Canals*, Oxford, Oakwood Press (2001)

Boyd, J.I.C., *Narrow Gauge Railways in Mid Wales (1850-1970)*, Oakwood Press, Oxford (1986)

Boyd, J.I.C., *The Festiniog Railway*, Odhams Press, London (1962)

Boyd, J.I.C., *The Wrexham, Mold and Connah's Quay Railway*, Oxford, Oakwood Press (1991)

Brindle, S., *Brunel: the Man who Built the World*, Weidenfeld & Nicolson, London (2005)

Brooke, D., *The Railway Navvy*, David and Charles, Newton Abbot (1983)

Brown, D.J., *Bridge: Three Thousand years of Defying Nature*, Octopus, London (2005)

Brown, R.A. *et al.*, *The history of the King's works, Volume 1: The Middle Ages*, HMSO, London (1963)

Buchanan, R.A., *Brunel*, Hambledon, London (2002)

The Engineers, Jessica Kingsley, London (1989)

Burton, A., *The Canal Builders*, Eyre Methuen, London (1972)

The Railway Builders, John Murray, London (1992)

Carter, E., *An Historical Geography of the Railways of the British Isles*, Cassell, London (1959)

Cartwright, R. and Russell, R.T., *The Welshpool and Llanfair Light Railway*, David and Charles, Newton Abbot (1989)

Chrimes, M.M., *Civil Engineering, 1839-89: A Photographic History*, Alan Sutton, Gloucester (1991)

Christiansen, R. and Miller, R.W., *The Cambrian Railways*, David and Charles, Newton Abbot (1967)

Colvin, H.M., A *Biographical Dictionary of British Architects*, 4th edn, Yale University Press, New Haven (2008)

Cross-Rudkin, P.S.M. and Chrimes, M.M., *Biographical Dictionary of Civil Engineers, 1830-1890*, TTL, London (2008)

Crow, A., *Bridges of the River Wye*, Lapridge Publications (1995)

Davies, H., *From Trackways to Motorways: 5000 Years of Highway History*, Tempus, Stroud (2006)

Davies, W.L., *Bridges of Merthyr Tydfil*, Cardiff, Glamorgan Records Office (1992)

Fowler, C.E., *The Ideals of Engineering Architecture*, Spon, London (1929)

Gorvett, D., *Bridges over the River Wye*, C.J. and A. Bryant, Whitney (1984)

Guy, A. and Rees, J., *Early railways 1*, London, Newcomen Society (2001)

Hadfield, C., *The Canals of South Wales and the Border*, David and Charles, Newton Abbot (1967)

Hadfield, C., *The Canals of the West Midlands*, David and Charles, Newton Abbot (1985)

Hadfield, C. and Skempton, A.W., *William Jessop, Engineer*, David and Charles, Newton Abbot (1979)

Hadley, B.M. and Dingwall, R., *The Wye Valley Railway*, Oakwood Press (1982)

Hague, D.B.R., *Lighthouses of Wales*, Royal Commission on Ancient and Historical Monuments in Wales, Aberystwyth (1994)

Helps , Sir A., *Life and labours of Mr Brassey*, Evelyn Adams and Mackay, London (1969, first published 1872).

Heyman, J., *The Masonry Arch*, Ellis Horwood, Chichester (1982)

Holden, J.S., *The Manchester & Milford Railway*, 2nd edn, Usk, Oakwood Press (2007)
Hopkins, H.J., *A Span of Bridges*, David and Charles, Newton Abbot (1970)
Hughes, S., *The Archaeology of the Montgomeryshire Canal*, Royal Commission on Ancient and Historical Monuments in Wales, Aberystwyth (1988)
Hughes, S. , *The Archaeology of an Early Railway System, Brecon Forest Tramroad*, RCAHMW, Aberystwyth (1990)
James, B., *G.T. Clark: Scholar Ironmaster in the Victorian Age*, Cardiff, University of Wales Press (1998)
Jervoise, E., *The Ancient Bridges of Wales and Western England*, Architectural Press, London (1936); republished by E.P. Publishing, Wakefield (1976)
Jones, S.K., *Brunel in South Wales, Volume I: In Trevithick's Tracks (2005); Volume II: Communications and Coal (2006); Volume III: Links with Leviathans (2009)*, Tempus Publishing Ltd, Stroud
Brunel in South Wales, Tempus Publishing Ltd, vols 1 & 2 (2005-6)
Jones, T., *A History of the County of Brecknock* (1909)
Lewis, M.J.J., *Early Wooden Railways*, London, RKP (1970)
Macdermot, E. T. (revised Clinker, C.R.), *History of the Great Western Railway, Volume 1: 1833-1863; Volume 2: 1863-1921*, Ian Allan, London (1982)
Margary, I.D., *Roman Roads in Britain*, John Baker, London (1973)
Marshall, J., *A Biographical Dictionary of Railway Engineers*, David and Charles, Newton Abbot (1978)
Nelson , J., *LNWR Portrayed*, Peco Publications, Seaton, Devon (1975)
Otter, R.A., *Civil Engineering Heritage: Southern England*, Thomas Telford, London (1994)
Popplewell, L., *A Gazetteer of the Railway Contractors and Engineers of Wales and the Borders*, Mellegden Press (1984)
Quartermaine, J., Trinder, B. and Turner, R., *Thomas Telford's Holyhead Road*, Council for British Archaeology, York (2003)
Penfold, A., (ed.), *Thomas Telford: engineer*, Thomas Telford, London (1980)
Phillips, D.R., *History of the Vale of Neath*, self-published (1925)
Place, G., *The Rise and Fall of Parkgate's Passenger Port for Ireland*, Carnegie, Preston (1994)
Price, M.R.C., *The Llanelly & Mynydd Mawr Railway*, Oxford, Oakwood Press (1992)
Pugsley, Sir A. (ed.), *The works of Isambard Kingdom Brunel*, Institution of Civil Engineers and University of Bristol (1976)
Rapley, J., *The Britannia and other Tubular Bridges*, Tempus, Stroud (2003)
Rattenbury, G., *Tramroads of the Brecknock & Abergavenny Canal*, Oakham, RCHS (1980)
Reader, W.J., *Macadam: The McAdam Family and the Turnpike Roads, 1798-1861*, Heinemann, London (1980)
Rennison, R.W., *Civil Engineering Heritage; Northern England*, Thomas Telford, London (1996)
Rolt, L.T.C., *George and Robert Stephenson*, Penguin Books, Harmondsworth (1978)
Isambard Kingdom Brunel, Penguin Books, Harmondsworth (1989)
Railway adventure, David and Charles, Newton Abbot (1970)
Thomas Telford, Penguin Books, Harmondsworth (1979)
Rowson, S. and Wright, I.A., *The Glamorganshire and Aberdare Canals* (two vols) (2001)
Ruddock, E.C., *Arch Bridges and their Builders, 1735-1835*, Cambridge University Press, (1979)
Russell, R., *Lost Canals of England and Wales*, David and Charles, Newton Abbot (1971)
Saint, A., *Architect and Engineer: a Study in Sibling Rivalry*, Yale, New Haven (2007)
Schnitter, N.J., *A History of Dams*, Balkema, Rotterdam (1994)
Simmons, J. and Biddle, G., *The Oxford Companion to British Railway History*, Oxford, University Press (1997)
Skempton, A.W., *British Civil Engineering, 1640-1840: A Bibliography of Contemporary Printed Reports, Plans and Books*, Mansell Publishing, London (1967)
Skempton, A.W., *Biographical Dictionary of Civil Engineers, 1500-1830*, TTL, London (2002)
Smiles, S., *The Lives of the Engineer*, Murray, London (1862) (and subsequent editions)
Spencer-Silver, P., *Tower Bridge to Babylon: the Life and Work of Sir John Jackson*, Six Martlets, Sudbury (2005)
Steel, W.L., *The History of the London and North Western Railway*, Railway and Travel Monthly, London (1914)
Stevenson, D.A., *The World's Lighthouses before 1820*, Oxford University Press (1959)
Telford, T. (ed. Rickman, J.), *Life of Thomas Telford, with a folio atlas of copper plates*, London (1838)
Troyana, L.F., *Bridge Engineering*, TTL, London (2003)
Turner, K., *Cliff Railways of the British Isles* (2002)
Turnock, D. A., *Historical Geography of Railways in Great Britain and Ireland*, Ashgate, Aldershot (1998)
West, G., *The Technical Developments of Roads in Britain*, Ashgate, Aldershot (2000)
Whishaw, F., *The Railways of Great Britain and Ireland*, Weale, London (1842, reprinted by David and Charles, Newton Abbot (1969)
Yeomans, D.T., *The Trussed Roof: its History and Development*, Scolar Press, Aldershot (1992)

NAMES INDEX

PLACES INDEX

NB: Welsh names are indexed under the English translation where appropriate; e.g. 'Pont' under 'bridge' or 'viaduct'; 'Afon' under 'river'; illustrations are listed in **bold** type; places are listed under modern spelling, e.g. 'Llanelli' for 'Llanelly'.